BROOKLANDS C
HEATH ROAD, WEYBR
Tel: (019
D1147413

EXTRAORDINARY WEATHER

Brooklands College
Library
WITHDRAWN FROM STOCK

EXTRAORDINARY WEATHER

Wonders of the atmosphere from dust storms to lightning strikes

Richard Hamblyn

CONTENTS

FOREWORD

The weather is a very British obsession. In fact, I'd say that's an understatement. The 'great British weather', as it's known to many in this country, is part of the fabric of our nation – a natural force that binds us together. Whether it's "too hot", "too cold", "too wet", or "too windy", even in the middle of a heatwave the weather breaks the ice in conversations.

As the striking photographs in this book attest, the weather in all its guises equally fascinates people around the world. From the simple act of raising an umbrella, to an aircraft that changes course to avoid a thunderstorm, every day the weather touches all our lives. I know many readers share my interest in the weather, as the ever-growing band of dedicated observers and amateur forecasters shows. But such is my passion for the weather I've made it my career. Over the course of nearly 30 years at the Met Office – the UK's national weather service – I've been captivated by the weather's power and beauty. Through forecasting, I've come to understand better its whims and to appreciate the magic I see day by day just by looking up at the sky.

Today, of course, all of us wherever we are, face the challenge of climate change, which is already bringing more extreme weather, more frequently to parts of the world. But whenever the weather makes its presence felt, we at the Met Office have an eye on the sky: from advising the British public on whether to expect rain as the occasional drop, drizzle or downpour, through to briefing multinational corporations on what the future climate may bring. To me, all weather is extraordinary for its ability to unite people of all ages and walks of life, even when it's "raining cats and dogs".

Rob Varley
Met Office Operations and Services Director

INTRODUCTION

The wayward beauty of our atmosphere flows from its essential dynamism, from the fact that there is never a moment when nothing can be said to be happening. At any one time, more than a thousand thunderstorms are in progress around the world, their restless summits rising towards the stratosphere, 10km (6 miles) or more above the Earth. Somewhere right now a tornado will be spinning, lightning crackling across the sky, and hail clattering on to roofs; elsewhere a dust storm will be advancing across a desert while, somewhere else, the weight of rime ice will be bringing down a power line. Meanwhile, perhaps, someone on the other side of the world will be photographing a miraculous sunset.

'The sky is the ultimate art gallery above,' as Ralph Waldo Emerson famously remarked – an open-air gallery that is perpetually rearranging its displays, from the great summer spectaculars of lightning and hailstones to the quieter, more contemplative visions of rainbows, haloes and crepuscular rays, all of which, and much else besides, can be found in the six themed chapters of this book: Storms and Tempests; Ice and Snow; Heat and Drought; Atmospherics; Strange Phenomena; and Man-made Weather.

Extraordinary Weather has been written with the view that all types of weather are extraordinary, but some are more extraordinary than others. Take the eerie display of mammatus clouds on page 121, for example, which loom towards Earth as though part of some imminent alien invasion, or the 'double trouble' storm image on page 31, in which a waterspout and a lightning strike make a brief elemental combination. These images seem so strange and otherworldly, and have such intense dramatic appeal, that it is easy to forget that, astonishing as they are, the events they depict are thoroughly normal. Indeed none of the weather moments that feature in this book could be categorized as 'freakish' or 'supernatural' at all – which is not to say that all of them are natural. As was the case with my earlier collection, *Extraordinary Clouds* (2009), one of the most eye-opening sections of this book is the one devoted to the subject of man-made weather, from artificially triggered lightning to contrails and ship tracks: this is a category of weather that has become increasingly prevalent in recent years, and will go on to be more prevalent in the future.

We have known for some time that human activity is altering the climate, but to what extent is it altering the weather? This is a complex question to which no adequate answers can yet be given, but we can certainly look at the cloud layer above densely populated regions and see that aircraft contrails are now among the most abundant cloud types on the planet (see page 133 for a satellite view of contrails above the southeastern United States). Ship tracks, meanwhile, paint a similarly crowded picture over the world's ocean highways, being bright lanes of cloud that form around the aerosol particulates from the smokestacks of ocean-going vessels (page 134). Recent research into fallstreak holes suggests that these, too, are largely anthropogenic effects, created by the curious physics of heavier-than-air flight: the expansion of a parcel of cloud-laden air as it passes over an aircraft's wings causes that air to cool, and such momentary cooling can be enough to prompt the supercooled droplets in the cloud to freeze and fall away, leaving a series of characteristic 'holes' in the cloud (see pages 104 and 135). As many of the images in the book confirm, human actions are dramatically altering the appearance of skies across the world, and even – in the case of noctilucent clouds – out towards the fringes of space (see page 138).

But it is not just the look of our skies that is changing. In May 2011, the charity Oxfam published a report entitled *Time's Bitter Flood*, which pointed to the fact that while the frequency of geophysical events such as earthquakes and volcanic eruptions has remained more or less constant over recent years, disasters associated with flooding and storms have risen from around 130 a year in the 1980s to more than 350 a year today: an increase of more than 4 per cent per year. In 2011 alone, there were record-breaking floods in Australia (where, in the country's worst natural disaster, an area the size of Germany and France remained inundated for two months), Brazil (which also experienced the worst floods in its history) and the American state of Missouri, where large sections of the Missouri River burst its banks following record snowfall in the Rocky Mountains along with near-record rainfall in Montana. A record-breaking year for tornadoes, meanwhile, saw more than 1,700 in the United States alone, one of which, on May 22, 2011, killed 159 people in the city of Joplin, Missouri. In Mexico, the worst wildfires in the country's history coincided with an April heatwave that saw temperatures peak at 48.8°C (119.8°F), while an unprecedented dry spell over western Europe saw wildfires breaking out in areas as unlikely as Ireland and the Highlands of Scotland. Journalists around the world took to describing this catalogue of unruly weather as 'global weirding', but the reality is that deeper snowfalls, more widespread flooding, heavier rains and greater extremes of heat and cold now constitute the 'new normal', and are inescapable features of the ongoing climatic reconfiguration to which the world must learn to adapt.

Earth's atmosphere is of course a globally circulating system, but because most of us experience it on a local level, it is hard to imagine our day-to-day weather as part of a wider picture. Satellite imagery offers a powerful corrective to this, which is why each chapter of this book includes a number of satellite pictures that allow us to zoom out from an individual weather event to show it in its fuller context. The 'cloud shadows' image on page 83 is an instructive example, for not only does it demonstrate the mechanism of crepuscular ray formation, it also shows how their radial appearance is a trick of linear perspective, since from this altitude the shadows can be seen to be parallel. It is one of my favourite images in the book, not least because it reminds us that we cannot rely on first impressions when it comes to the turbulent world of weather, where things are not always what they seem. Turn, for instance, to the picture on page 110. Is it a fish? Is it a dragon? No, it's just an unusual side effect of the physics of cloud convection, but, in common with most of the other images I have chosen for this book, it has a powerful, if transient, ambiguity. Welcome to the world of extraordinary weather!

EXTRAORDINARY WEATHER
STORMS AND TEMPESTS

At any one moment more than a thousand thunderstorms are raging on Earth; here are just a few of them, along with their associated effects, including waterspouts, tornadoes and floods.

Cumulonimbus anvil

Cumulonimbus clouds are enormous structures that can grow as far as the lower stratosphere, around 20km (12 miles) above the Earth. The temperature inversion at the tropopause (the boundary between the troposphere and the stratosphere) causes the icy summits of these convective giants to spread horizontally, creating the characteristic sloping anvil shape, as seen from below in this close-up aerial photograph taken in August 1971 from a research plane of the United States National Oceanic and Atmospheric Administration (NOAA). Most aircraft pilots give cumulonimbus clouds a wide berth, given the strength of the convective currents inside them, as well as their high water content, which can leave thick ice deposits on cold metal. However, NOAA reconnaissance crews are on a mission to get close to some of the most turbulent clouds on Earth – indeed, as we will see on page 28, to actually get inside them.

Thunderstorms over Brazil

At any given moment, there are around 1,800 thunderstorms in progress somewhere on Earth. This dramatic photograph, taken by an astronaut on board a Space Shuttle in February 1984, shows just a few of them, massing over the Paraná River basin in southern Brazil. With abundant warmth and moisture-laden air in this part of the world, vigorous thunderstorms are commonplace, as are flash floods caused by heavy downpours of rain. These towering thunderheads exhibit characteristic flattened anvils, where the fast-rising clouds begin to spread horizontally as they reach the lowest part of the stratosphere, around 20km (12 miles) above the Earth; a number of overshooting tops can be seen pushing through the thermal boundary – an indication that these storms will be severe.

Cumulonimbus over West Africa

This stunning sideways view of a vast cumulonimbus anvil was photographed by an astronaut on board the International Space Station as it passed over West Africa, near the Senegal–Mali border, on February 5, 2008. The fully formed anvil cloud has numerous smaller cumulonimbus towers rising around it, a sure sign of the ongoing thermal activity that is filling the atmosphere with hazardous energy. Storm systems such as this bring torrential rain, lightning, ferocious winds and possibly tornadoes, fuelled by the energy constantly released by the vast quantities of water vapour that rise and rapidly condense inside the turbulent, unstable cloud mass. At the height of its power, a thundercloud of this magnitude will release the energy equivalent of an atomic bomb every couple of minutes.

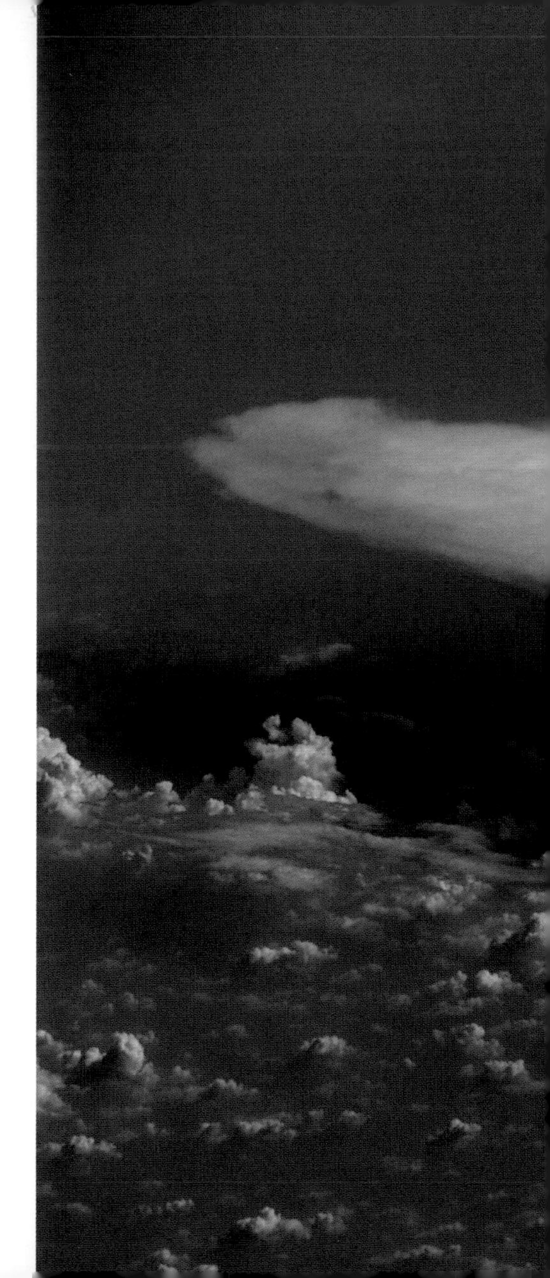

Gust front clouds

A gust front – also known as an 'outflow boundary' – is a leading edge of cold air that rushes downwards from a thunderstorm, spreading out fast when it hits the ground. Gust fronts are often marked by turbulent areas of cloud and icy rain, as in this example, churning above a field of wheat near the Wichita Mountains of southwest Oklahoma, USA, during a bout of severe thunderstorms in early May. As with other downdraught phenomena (see page 122), the wind shear associated with strong gust fronts can be extremely hazardous to aircraft, especially during takeoff and landing.

Inflow bands

Inflow bands (sometimes known as 'feeder bands') are streaks of low cumulus cloud that extend, parallel to the wind, from a thunderstorm's main tower; they show that a storm is drawing in low-level air from several kilometres away, and their appearance can give some indication of the potential severity of the storm. If the inflow bands appear as spirals (see next page for a dramatic example), then cyclonic rotation has begun, and the storm may turn tornadic. In this example, however, photographed in northwest Texas, USA, one June afternoon, the bands stream radially into the storm cloud; there is lightning, hail and heavy rain, but – so far – little chance of a tornado.

Sunset supercell, Iowa

Resembling a scene from the film *Independence Day*, a wind-sculpted supercell storm cloud flails above a highway near Sioux City, Iowa, USA, in May 2004. 'The storm was sucking inflow in extremely hard and creating some crazy structures,' recalls the photographer Mike Hollingshead. 'Note how smooth the cloud base is away from the centre – the speed of the inflow being sucked into that base must be incredible.' Hollingshead describes how an arm of cloud seemed to wrap itself around the base of the storm, creating a large coiled area: 'You can see a little sculpted lowering at the base and it looks like it might produce a tornado, but it doesn't.' Rain falling from the centre of the storm blocked the evening sun, and 'as the sun lowers to below the cell, I get this shot – one of my all-time favourites.'

Supercell lightning strike

'June 5 has frequently been a lucky day for me,' recalls noted storm chaser Jim Reed, and that was especially true of June 5, 2004, which produced this, the best-known image of his career. Having spent much of that day chasing thunderstorms across southern Kansas, USA, the sun was just about to set when Jim set up his tripod by the side of the road in order to photograph this isolated, saucer-shaped supercell, backlit by the setting sun. 'On the second squeeze of my shutter,' he writes, 'a lone, magenta-coloured lightning bolt dropped from the storm: "a raw scorch of lightning". . . The image represents everything I love about storm chasing . . . serendipity, technology, art, and atmospheric magic.'

Wall cloud, Oklahoma

A regular feature of supercell storm systems, wall clouds (sometimes known as 'pedestal clouds') are isolated lowerings attached to the bases of storm clouds. They develop when rain-cooled air is pulled towards the mesocyclone (the supercell's violently rotating core), its moisture condensing at a lower level than the principal cloud itself. Wall clouds thus mark the area of strongest updraught within the storm, and are characterized by extremely strong winds. Tornadoes often develop from descending wall clouds, as can be seen happening in this photograph, taken in Washita County, Oklahoma, USA, in June 1980. A dark band of heavy rain is following close behind.

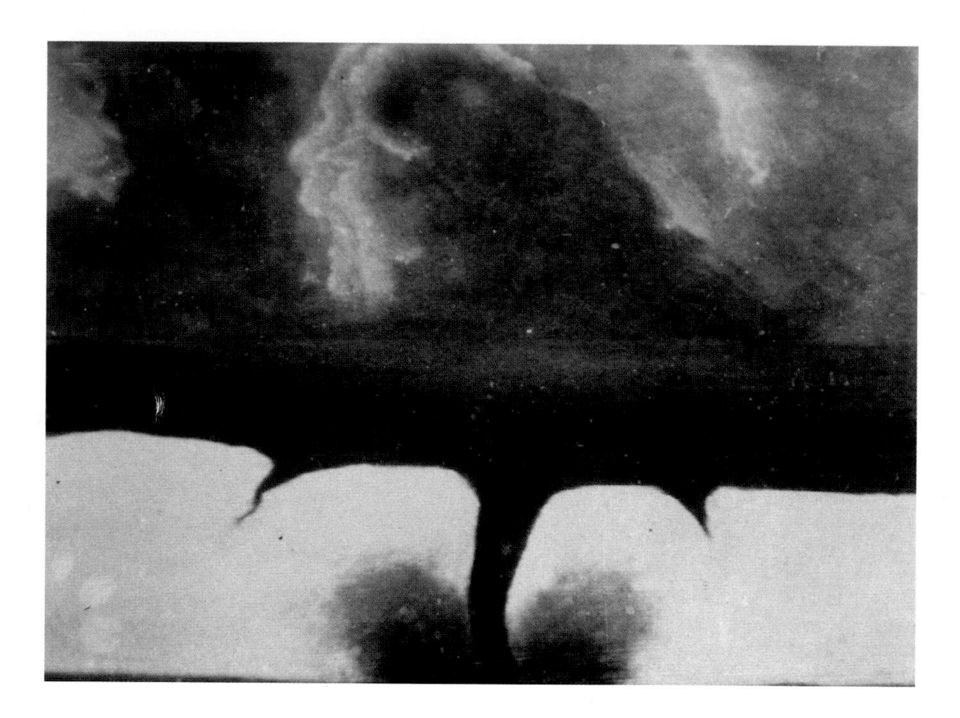

Nineteenth-century tornado

'The populous region of the United States is forever doomed to the devastation of the tornado,' observed the American meteorologist and writer John Park Finley in 1887. 'As certain as that night follows day is the coming of the funnel-shaped cloud.' Tornadoes were always a feature of American frontier experience, but it was the advent of photography in the middle of the 19th century that made them familiar around the world. There were many technical difficulties involved in photographing skies on long-exposure plates, but early photographers took some impressive tornado pictures, such as this – one of the earliest known photographs of a tornado – caught by F. N. Robertson of Howard City, South Dakota, on August 28, 1884. It's a powerful image: the main funnel kicks up a dust cloud on the ground, while a pair of secondary funnels emerge like horns on either side, giving the image a faintly devilish air.

Tornado with lightning

A tornado is a violent, rotating column of air that connects the base of a supercell storm cloud to the ground below. Tornadoes form when powerful downdraughts drag the supercell's mesocyclone (the rotating central updraught) towards Earth, generating a visible condensation funnel that begins to throw up soil and debris as soon as it makes contact. In this dramatic example, large amounts of red topsoil from a wheat field in the American Midwest are being hurled around (and inside) the base of the funnel, as cloud-to-ground lightning leaps from the rain-soaked thundercloud behind.

Death of a small town

This close-up shot shows debris flying from the base of the wedge tornado, 1km (1,100yd) wide, that battered the small settlement of Manchester, South Dakota, USA, on June 24, 2003 (a date remembered locally as 'Tornado Tuesday' because of the large number of powerful twisters that were generated that day). Winds of up to 320km/h (200mph) destroyed every building in the century-old Manchester community, which was subsequently abandoned and never rebuilt. A lone granite memorial now marks the spot where the ill-fated town once stood.

Roping out tornado

A tornado can last from a few seconds to more than an hour, but at some point the energy that sustains it will ebb and the tornado will begin to 'rope out' before the rotating funnel fragments and disappears. Tornadoes remain highly dangerous at the rope stage, especially if they are being blown around by the wind. The one in this photograph, taken near Cordell, Oklahoma, USA, on May 22, 1981, was known locally as the 'Wizard of Oz tornado' because of its unusually slender shape. NOAA footage of the twister, available on YouTube, shows how vigorous its final moments were, as it roped out into a sinuous curve before dissipating. It was filmed as part of the 'Sound Chase' project, jointly run by the National Severe Storms Laboratory and Mississippi State University, which sought to monitor the various noises made by tornadoes at ground level.

The eye of the storm

This amazing photograph of the eye of Typhoon Nabi was taken from the International Space Station on September 3, 2005; by then the typhoon was four days old, and at the height of its powers as it tracked across the western Pacific towards South Korea and Japan. The sharply defined eye marks the region of lowest pressure within the storm, and is an area of relative calm surrounded by the ferocious winds of the eyewall (see next page). A typhoon is essentially the same phenomenon as a hurricane or a tropical cyclone – a powerful storm system that develops around a large low-pressure centre. Their different names are related to their geographical locations: 'typhoon' describes a severe tropical storm over the western Pacific; over the Indian Ocean such a storm is called a 'cyclone'; and over the eastern Pacific and Atlantic oceans, a 'hurricane'. (If an eastern Pacific hurricane tracks into the western Pacific, it is re-designated as a typhoon).

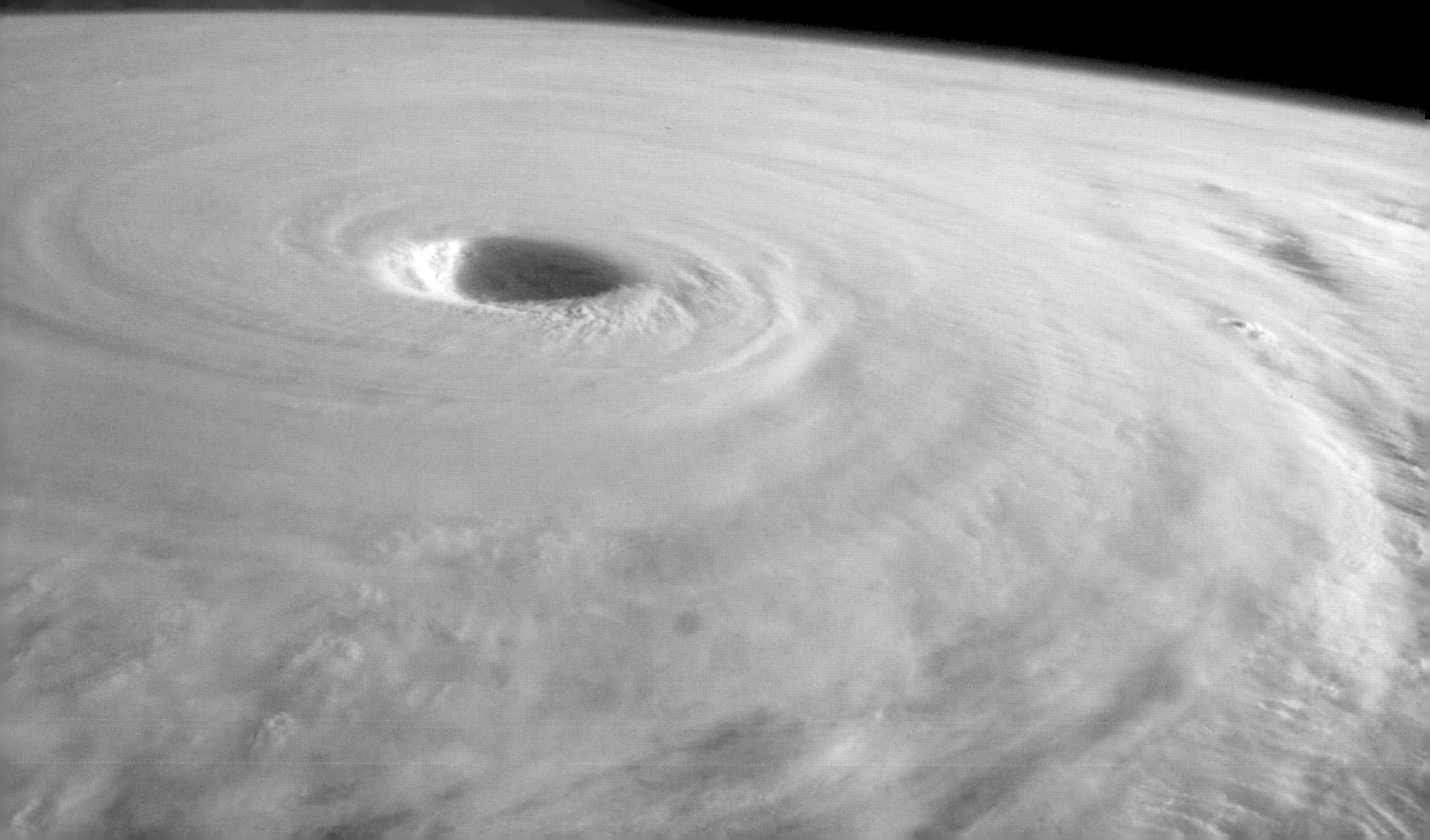

Inside the hurricane

The calm eye of a hurricane (and of a typhoon or a tropical cyclone) is a roughly circular area 30–65km (20–40 miles) in diameter. It is surrounded by the eyewall, a ring of towering thunderstorms where the strongest winds and most severe weather are to be found. As mentioned on page 12, NOAA sends research planes into the teeth of severe storms and hurricanes in order to collect information that will determine their likely strength and direction. I once interviewed a retired weather pilot who had spent his career flying directly into hurricanes and supercell storms. When I asked him what kind of training he had received for such missions, he laughed and said: 'Your on-the-job training is accomplished the first time you fly into one.'

Storm of war

Waterspouts are essentially tornadoes at sea, though they tend to be weaker and less destructive than their land-based counterparts. The name 'waterspout' is somewhat misleading, because – as with tornadoes – their spouts consist of atmospheric cloud condensed within the whirling funnel, not seawater picked up from the surface. They can grow to an impressive size, like this unusually powerful tornadic spout, which was photographed from an aircraft accompanying a North Atlantic convoy during World War II. Its cascade (the plume of spray whipped up from the surface) appears to dwarf the merchant ship that steams nervously past it at a distance of less than 1km (1,100yd).

Double trouble

This extraordinary night-time scene was photographed near Lake Okeechobee in Florida, USA, during a severe thunderstorm in June 1993. At the height of the storm – around ten in the evening – a funnel cloud descended from the base of the storm cloud, connecting with the lake to form a powerful waterspout. The lucky amateur photographer, Fred Smith, was standing in his backyard when the lightning struck, illuminating the monster spout that towered over the 150m (490ft) television mast to the right of the image. Though such inland waterspouts are usually fairly short-lived and rarely cause significant damage, they can be an impressive (and noisy) sight, as this amazing photograph suggests.

Floods in southeast Texas

During the night of May 28–29, 2006, an unusually long-lived thunderstorm delivered some 410mm (16in) of rain to southeast Texas, USA, leading to widespread flooding along the coastal plain. The following afternoon, a break in the clouds allowed NASA's Aqua satellite to capture this Moderate Resolution Imaging Spectroradiometer (MODIS) image, in which the extent of the flooding can be seen at a glance. Water, which appears blue/black in this colour visualization, forms dark pools east of Trinity Bay; to the west, the city of Houston spreads like a vast skein of concrete, pale grey against the drowned vegetation.

Tornado wasteland

On April 26, 1991, an outbreak of 55 separate tornadoes brought death and destruction to large areas of Kansas and Nebraska, USA. Tornadoes are classified from 0–5 on the Fujita Scale, depending on wind speed and associated damage. The city of Andover, Kansas, was hit by a F5 tornado 1km (1,100yd) wide, which destroyed more than 300 homes before moving northeast towards the Golden Spur Mobile Home Park; as this photograph shows, the entire trailer park was reduced to shattered debris by the tornado's 320km/h (200mph) winds. Thirteen people were killed at the site, as were many more elsewhere that day. Even so, the outbreak was not the most deadly of recent times: the so-called Super Outbreak of April 3–4, 1974, spawned 148 confirmed tornadoes across the eastern United States, which claimed 319 lives, while the tornado that hit the town of Joplin, Missouri, on May 22, 2011, killed 159. Though it is not easy to say whether tornadoes and other weather disasters have become more frequent in recent years, they have certainly become more costly. During the 1980s, severe weather events cost the insurance industry an estimated $25 billion a year, whereas today they cost insurers around $130 billion a year.

EXTRAORDINARY WEATHER

ICE AND SNOW

From blizzards and ice storms to fragile snow bales, this section looks at some of the many curious effects that are created by snow and ice.

Frozen Britain

Britain froze to a standstill during the first week of January 2010, after deep snow blanketed much of the British Isles, and the coldest temperatures in 30 years turned roads and pavements into virtual ice rinks. Heavy snowfall damaged power lines in southern England, leaving several thousand homes without electricity, as subzero air from the Arctic and Scandinavia brought overnight temperatures as low as –18°C (–0.4°F) in isolated areas. This extraordinary image, captured by NASA's Terra satellite on January 7, 2010, made the front pages of most British newspapers the following day, beneath headlines such as 'Frozen Britain' and 'UK as cold as the South Pole'.

Ice cars

One January morning in 2006, strong winds blew surface water from Lake Geneva over parts of the lakeside town of Versoix in Switzerland. The air temperature was extremely cold, around –6°C (21°F), and the fast-moving spray froze on every surface it touched, forming thick layers of icicles that covered vehicles, roads and trees. The photogenic aftermath was widely misreported as being the result of an ice storm, but this was no ice storm, merely the action of wind-blown water hitting already frozen objects at high speed. True ice storms, by contrast, occur when a layer of warm air is sandwiched between layers of cold air. Snow falling from the cold upper layer melts in the warmer section, then refreezes to form treacherous glaze ice as soon as it hits the ground (see page 39).

Frost feathers

Thick deposits of rime ice coat the pillars of a meteorological station in the Swiss Alps, following a high-altitude winter storm. Rime ice forms when tiny droplets of supercooled water come into contact with a frozen surface (supercooled droplets remain in a liquid or semi-liquid state even at subzero temperatures). Remarkably thick deposits of rime can form on the windward side of an object, sometimes building up into large, tabular features known as frost feathers. Rime ice such as this is normally found only at higher altitudes, where supercooled fog or drizzle persists in the cold, clean air.

Ice storm

As mentioned on page 37, an ice storm occurs when a layer of warm air becomes sandwiched between layers of cold air. Snow that falls from the cold upper layer melts when it passes through the warmer middle section, before refreezing into glaze ice as soon as it hits cold surfaces at ground level. The United States National Weather Service defines an ice storm as an event that results in the accumulation of at least 6.4mm (¼in) of ice on exposed surfaces: as can be seen in this undated photograph from NOAA's collection of historic weather images, the weight of such deposits can bring down trees and power lines, adding another dimension to winter woe.

Hail shower

This large cumulonimbus cloud over Exeter, Devon, England, produced a shower of sizable hailstones in early April 2008. Hail is formed when warm updraughts of air hurl descending ice pellets back up into the colder regions of the cloud. This can happen several times, during which the yo-yoing pellets grow concentrically through collision and freezing, creating ever bigger and weightier stones until gravity – aided by downbursts of cold, turbulent air – sends them flying to the ground.

Hail on the streets

Not the result of an overactive ice machine at this Midwest branch of 7-Eleven, but a barrage of hail that carpeted streets and sidewalks during a severe supercell thunderstorm on May 11, 1982. The storm went on to produce a category F2 tornado, but hail is as much a hazard as a twister: as can be seen in the foreground of this picture, taken by a member of NOAA's National Severe Storms Laboratory intercept team, the larger stones appear to be 5–8cm (2–3in) in diameter – about the size of a billiard ball – and are hitting the ground extremely hard. No wonder the highway is deserted.

Hail-scarred apples

Crops such as cereals and fruits are particularly vulnerable to hailstorms; the annual cost of hail damage to farmers in the United States has been estimated at $1.3 billion – 1–2 per cent of the total annual crop value. Australian apple growers suffered severe hailstorms during the (southern) summer of 2010, one of which, on November 19, caused losses valued at $2.6 million, destroyed a number of orchards, and was declared a natural disaster. Hail damage depends on the maturity of the fruit as well as on the size of the hailstones. A hard impact early in the growing season can cause a deep depression and deformation of the fruit, while damage later in the season can appear more bruise-like. The young apples in this picture are beyond recovery, their skins scarred and pitted by heavy hail.

The lake effect

Lake-effect snow is caused by cold Arctic air passing over a warm body of inland water. The cold, dry air picks up the lake's warmth and moisture and condenses the latter into snow-laden stratocumulus clouds, which can deposit large amounts of snowfall downwind from the lake. That can be seen happening in this photograph, taken from above the North American Great Lakes region in December 2000. This view of lakes Nipigon (top), Superior (middle), and Michigan (bottom) shows striking contrasts between clear and cloudy air as the cold wind blows from the northwest across the lakes, generating powerful snowstorms over Michigan and Ohio.

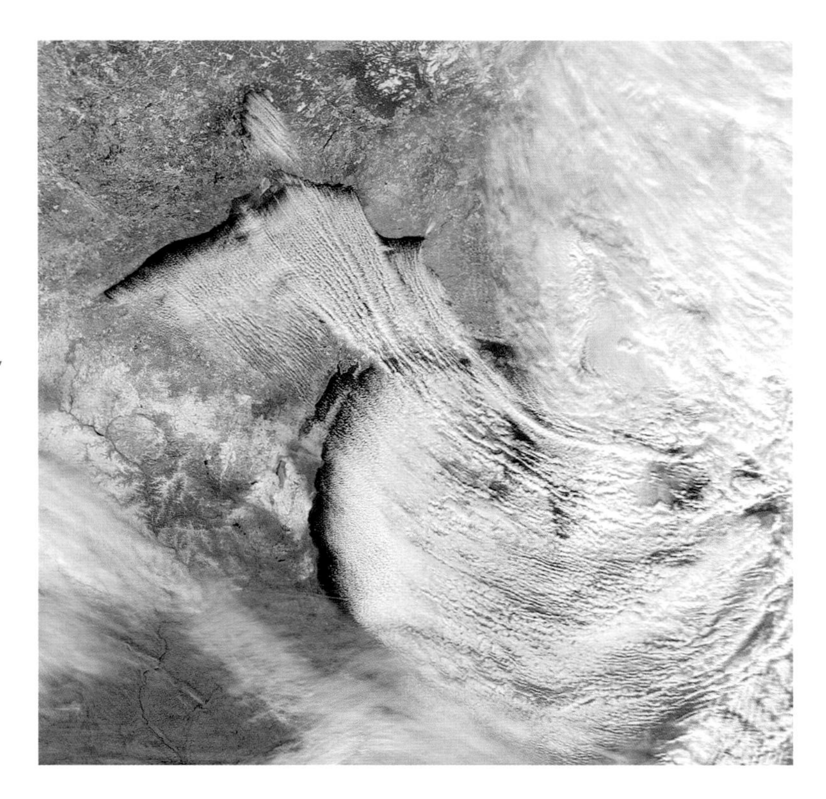

North Dakota snowdrift

On March 9, 1966, much of Jamestown, North Dakota, USA, lay buried under 4.5m (15ft) of snow, following an unusually powerful blizzard. (The United States National Weather Service classifies a winter storm as a blizzard if it contains large amounts of snow or blowing snow, if the wind speed is 56km/h (35mph) or more, the visibility is less than 400m (1,312ft) and if conditions persist for a minimum of three hours.) The photographer, Bill Koch, of the North Dakota State Highway Department, summed up this extraordinary event in a single potent image by positioning one of his colleagues next to an almost submerged utility pole. When this photograph first appeared in the local press, the caption read, 'I believe there is a train under here somewhere!' It's entirely possible that there was.

Snow rollers I

Also known as snow bales or snow doughnuts, snow rollers are cylindrical snowballs formed by strong winds blowing over snow-covered open ground. Their formation depends on a particular balance of air temperature, moisture, wind speed and snow depth. Ideally, the top layer of snow should be nearly at melting point, so that it becomes slightly wet and 'sticky'; this layer can be peeled off by the wind from the colder, more powdery snow beneath, and then rolled into characteristic bale formations, which are often hollowed out by wind action (as in this photograph). Snow rollers can travel for surprising distances, but usually end up resting against a piece of vegetation or stranded at the bottom of a hill.

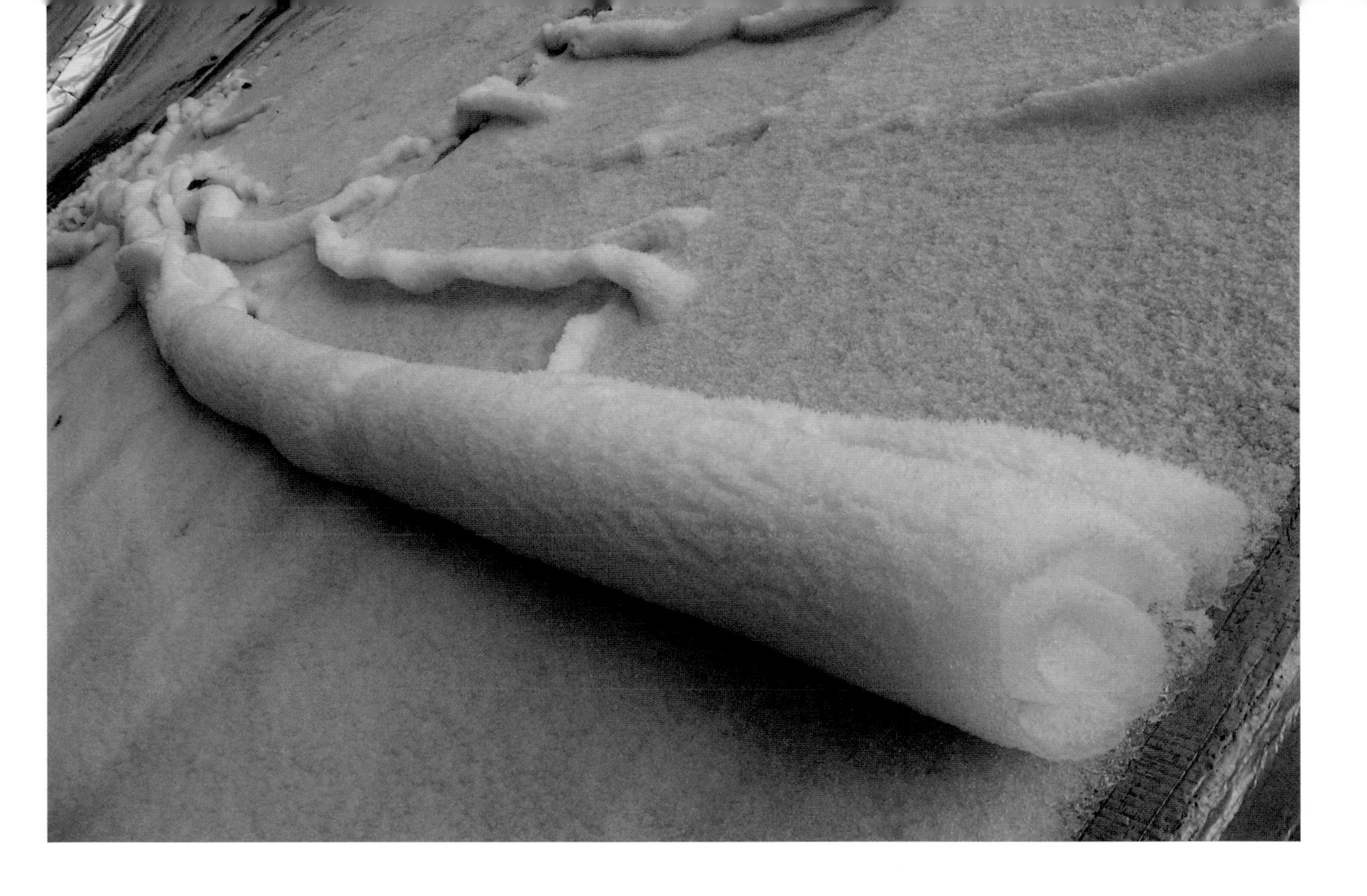

Snow rollers II

Snow rollers can also be formed solely by gravity, as in this unusual example, where a thin layer of sticky snow has rolled itself up like a carpet on the roof of a house in the mountainous Krkonoše district of southern Poland. The two different grades of snow that are required to create the phenomenon are clearly visible: the wet, loose snow that forms the roll itself, and the icy, powdery layer below, some of which can be seen sticking to the outside surface of the roll.

EXTRAORDINARY WEATHER

HEAT AND DROUGHT

Sandstorms, dust devils and raging wildfires: just a few of the heat-related phenomena with which we are becoming increasingly familiar as our atmosphere continues to warm.

Dust storm, Texas, 1935

The severe North American dust storms of the 'Dirty Thirties' were an early indication of humankind's capacity to influence weather and climate. Decades of intensive farming across the prairies of the United States and Canada had displaced the natural grasses, which had served to keep the soil in place and retain ground moisture even during extended dry periods. When severe drought afflicted the Great Plains in 1934, the overworked soil was easily eroded by storms blowing in from the west, and coalesced into suffocating clouds of turbulent, fast-moving particles. April 1935 saw some of the worst of these so-called 'black blizzards', such as the one seen here engulfing the town of Stratford, Texas, on April 18, 1935. By the late 1930s, President Franklin D. Roosevelt's soil conservation programme had done much to alleviate the situation. Today, however, there are many dryland areas in the developing world that are experiencing a similar increase in severe dust storms, as desertification spreads in the wake of climatic and agricultural change.

Sandstorm in Sudan

An intense sandstorm, known as a haboob (from the Arabic for 'strong wind'), advances on a livestock market in the city of Omdurman, Sudan, in November 2004. Relatively common in most arid regions, haboobs form very suddenly – often with no warning – when powerful downdraughts from nearby thunderstorms send vast billowing walls of dust and sand hurtling outwards at speeds of around 35–50km/h (20–30mph), covering everything in a layer of choking sediment.

Sandstorm from above

This amazing photograph was taken (using a digital camera with a 50mm lens) by an astronaut on board the International Space Station, while in orbit around the Earth on February 15, 2004. It shows a massive sandstorm sweeping like a silver tide over the Persian Gulf state of Qatar, heading south towards Saudi Arabia and the United Arab Emirates. A major low-pressure system over southwestern Asia had generated a series of such storms that scoured the region, but it is only from this heightened perspective that the magnitude of these events can be appreciated.

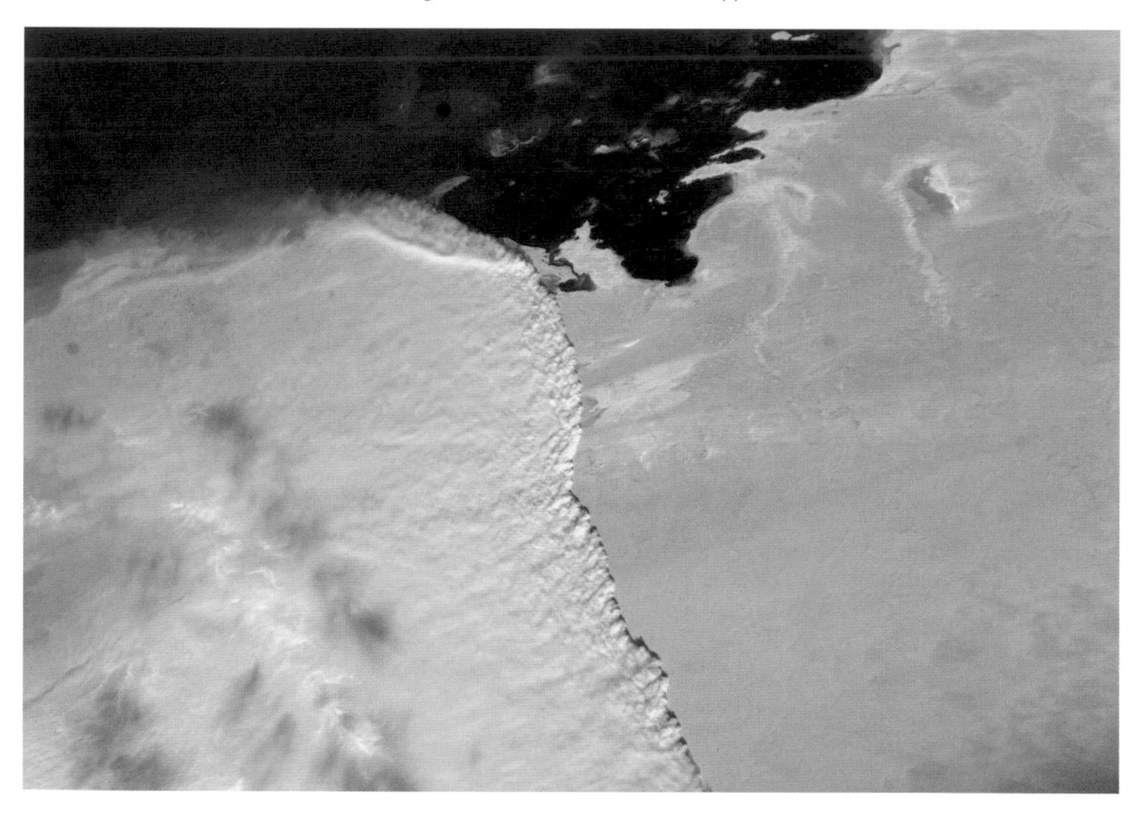

Dust cloud over the central Mediterranean

April 14, 1935, was known as 'Black Sunday' in the United States, for it was the day on which the worst of that year's Dust Bowl storms took place. Sixty-five years later to the day, on April 14, 2000, a vast dust plume billowed out over the central Mediterranean from an unusually intense North African sandstorm: the cloud of debris extended across the sea from Algeria to Italy, and was clearly visible from space. This date appears to have a strange geophysical significance, as it was on April 14, 2010, that the Icelandic volcano Eyjafjallajökull erupted, sending an ash cloud billowing out over much of northern Europe and grounding aircraft for more than a week (see page 108).

Sandstorm in western Iraq

As evening fell on April 27, 2005, a massive sandstorm rolled over western Iraq, enveloping much of the province of Al Anbar, including the Al Asad Airbase, one of the largest military installations in Iraq (and the former home of the United States' II Marine Expeditionary Force). These turbulent walls of dust and sand travel at surprising speeds, but a storm-driven haboob such as this is distinct from the notorious shamal, the northwesterly wind that blows across the Persian Gulf, which can create sizable long-distance sandstorms on a greater scale than this.

Dust devil

A vigorous dust devil sweeps across the San Joaquin Valley floor in central California, USA, on a warm August afternoon. Dust devils are formed in updraughts of air from localized heating in dry conditions, and begin to whirl when the warm wind is funnelled through an obstruction or over a patch of particularly rough ground. They vary in size from the smallest and most short-lived of wisps to intense dusty vortices, such as this one, capable of causing minor damage.

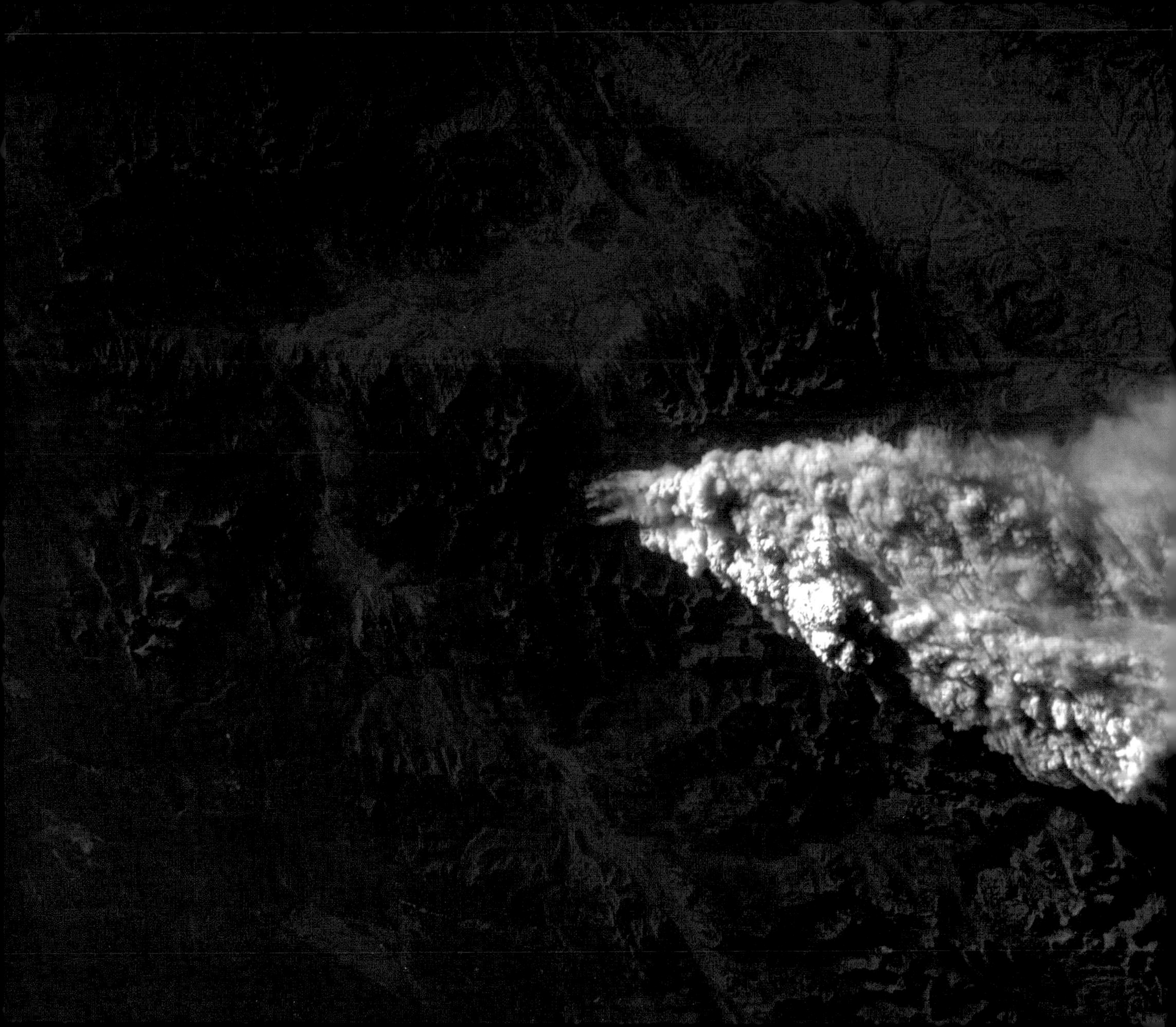

Wildfire, southern Montana

A wildfire rages in southern Montana, USA, its vast smoke plume visible from space: this photograph was taken in August 2007 by an astronaut working at the International Space Station, around 300km (186 miles) above the Earth. Wildfires are a natural part of most forest ecosystems, but they contribute to atmospheric warming by releasing significant quantities of greenhouse gases and particulates, and the problem is that they are becoming more frequent as the world continues to warm. Global wildfire predictions published by NASA in 2010 suggest that many of the world's forested areas will continue to experience bigger and more frequent fires, which in turn will contribute to further atmospheric warming, in a classic feedback loop. The areas that are likely to see the greatest increase in forest fires are India, Australia, Central Asia and Siberia, southern Europe, southern Africa and the dry western states of North America.

Fire whirl, Wyoming

On August 11, 2006, a dry thunderstorm blew through the city of Casper, Wyoming, USA, igniting several wildfires in the area, which had been subject to prolonged drought. One of these fires – the Jackson Canyon Fire – soon grew out of control, sending wide crescents of flame across the west side of Casper Mountain. The fire burned for several days, during which time a series of fire whirls were seen. Given the hazardous circumstances in which they appear, fire whirls are rarely photographed, but this dramatic image, shot by meteorologist Dan Borsum of NOAA's National Weather Service, shows how a rotating column of dry, heated air can detach itself from the main part of the fire, acquiring a vertical vorticity of its own. Depending on the air temperature and the strength of the wind, these burning whirlwinds can either remain within the burn area, dying down after a few seconds or minutes, or move away as separate entities, as eerie 'fire devils' or 'fire tornadoes', some of which can grow to impressive heights and last for 20 minutes or more.

Icelandic sandstorm

The Mýrdalssandur sand flat is a glacial outwash plain near the coast of southern Iceland – an extensive, uninhabited desert of fine alluvial sands and soils, left behind by glacial rivers and floods. This vast flat is subject to intense winds and gust fronts, and – as in any desert situation – sandstorms can form when enough dry, loose particles are picked up by the wind and transformed into a mobile column of grit. They tend to be short-lived but impressive events, as this wonderful photograph shows.

Haboob over Phoenix, Arizona

The kind of intense dust storm known as a haboob is a feature of the majority of arid landscapes, including the dryland regions of North America, which are experiencing an increase in such storms. On the evening of July 5, 2011, thunderstorms in eastern and southern Arizona collided, with the strong downdraughts and dry conditions combining to raise a dust storm 1km (3,300ft) high and nearly 160km (100 miles) wide, which roared through the city of Phoenix at nearly 100km/h (60mph). Visibility in the city was reduced to almost nothing as the storm passed through; in its wake it left a wide trail of dust, debris and mud-choked swimming pools. The 2011 Phoenix haboob was a bigger version of the one that enveloped the southern part of the city on August 22, 2003, as captured in this amazing picture.

EXTRAORDINARY WEATHER

ATMOSPHERICS

From Brocken spectres and swirling auroras to storm clouds lit by shards of lightning, the atmosphere is home to some of nature's greatest optical displays.

Seas of cloud

A blanket of fog fills a mountain valley at sunrise, beneath the orange glow of underlit altocumulus cloud. Such spectacles are a reminder that we live not only on the surface of the solid Earth, but also at the bottom of a turbulent ocean of air. Valley fog forms when a pocket of cold air settles into a low-lying area while a layer of warmer air passes over the hills above. These overnight fogs usually dissipate shortly after sunrise, but in calm, stable conditions they can persist for several days.

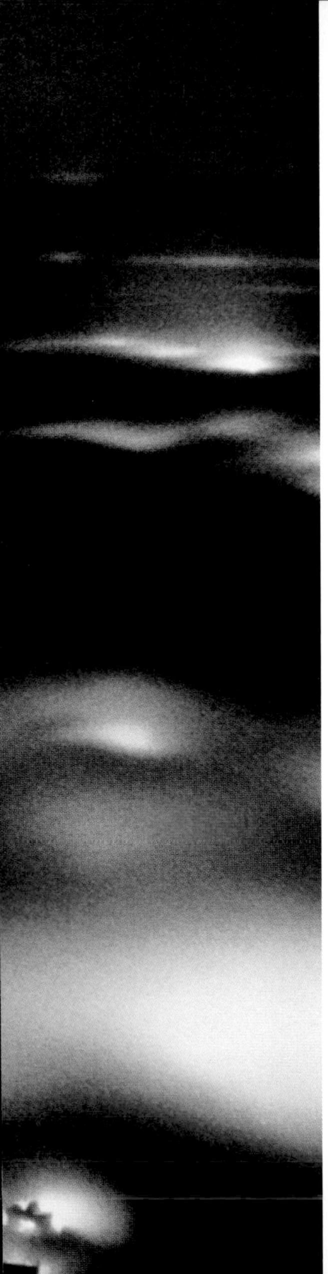

Glowing fog

This luminous landscape – seen from the Euganean Hills, just south of Padua, Italy – is the result of a variety of village street lighting diffusing through a layer of nocturnal radiation fog (formed by the cooling of the land after sunset). The different colour temperatures of the various forms of lighting make the fog glow in a range of colours. The yellow areas are generated by ordinary tungsten filament lighting, the red areas by sodium vapour lights and the blue ones by mercury vapour lights. It may look like Las Vegas by night, but in fact this is a rural scene – testament to the huge amounts of energy that the world so casually wastes.

Double lightning strike

Lightning is the visible result of the build-up of electrical charges within cumulonimbus thunderclouds. Clusters of positive and negative charges accumulate in different parts of the cloud, with positive charges often congregating at the top and negative charges at the base. When the potential difference between these charged areas becomes too great to sustain, electrical energy is discharged in the form of lightning. Though most lightning is in-cloud lightning (often referred to as 'sheet lightning' because it is seen only as a diffuse brightening of the cloud), cloud-to-ground (or 'forked') lightning is the more familiar type. It strikes when negative charges in the base of a cloud induce opposite charges in the ground below, as in this unusual display of twin lightning bolts bursting from a low shelf of eerily lit storm cloud.

Horizontal lightning

Cloud-to-cloud lightning, one of many kinds of lightning that do not reach the ground, is an electrical discharge that occurs between differently charged areas of cloud, either within a single cloud or between two separate clouds. They are often some distance apart, as with this dramatic horizontal stroke photographed above the town of Orange, New South Wales, Australia, by Shane Lear, a welder and keen amateur fulminologist (student of lightning). Mr Lear takes many of his lightning photographs from the roof of his house, where he has an array of six horizon-spanning cameras that can be activated simultaneously in order to capture some extraordinary electrical effects.

Anvil crawler

An 'anvil crawler' is a variety of multichannelled in-cloud (or cloud-to-cloud) lightning that appears to move around the underside of a thunderstorm's towering anvil. As its name suggests, it is a relatively slow kind of electrical discharge, and is usually associated with weak or decaying storm systems, as in this fine example seen snaking through a double rainbow in the wake of severe thunderstorms in northwestern Oklahoma, USA, on May 29, 2006.

Lightning over Budapest

Around midnight on August 28, 2001, a massive thunderstorm broke over the city of Budapest, Hungary. The weather was hot, as the photographer Roger Coulam remembers, and the storm had been brewing all day. 'I rushed out of bed as the storm began, to spend an amazing hour photographing in heavy rain. This massive positive lightning bolt over the lights of the sleeping city shows where the power really lies!' His striking image shows a number of weaker strokes crackling through the charged air, alongside the powerful white bolt that slams into the city: a direct hit from above.

Rocket lightning

Shortly after 2am on August 30, 1983, lightning struck the Space Shuttle *Challenger* as it prepared to launch from the Kennedy Space Center in Florida, USA. This photograph – which won the 1983 World Press Photo prize in the Arts and Sciences category – was taken by a remote camera set up by Sam Walton of United Press International. It captures the extraordinary power of the 'stepped' cloud-to-ground lightning stroke, as well as the torrential rain that was whipped up by the night-time thunderstorm. Amazingly, the launch was delayed by only 17 minutes, and the mission went ahead without further incident. Lightning is a familiar hazard at Kennedy Space Center launches: Apollo 12 was struck by lightning twice during liftoff in November 1969, causing circuits on the fuel cells to malfunction briefly.

Supercell with lightning strike

Cloud-to-ground lightning strikes the Earth during a supercell thunderstorm in southwest Oklahoma. A supercell is a severe thunderstorm that develops around a deep rotating updraught known as a mesocyclone (see page 21). Supercells tend to generate heavy rain, hail and lightning, as well as ferocious winds, which in this case reached nearly 130km/h (80mph). The smooth, curving shelf cloud at the base of the storm makes a dramatic contrast with the sharp downstroke of the lightning bolt that illuminates the rain-darkened horizon.

Rainbow above Victoria Falls

This photograph, taken in September 1989, shows a rainbow formed in the spray of the world's largest waterfall, Victoria Falls, on the Zimbabwe–Zambia border in southern Africa. The falls' local name, Mosi-oa-Tunya, which translates as 'the smoke that thunders', describes both the noise and the spray thrown up by the plunging waters as they fall some 108m (360ft) into the first of a series of gorges carved by the Zambezi River. Rainbows (or, at night, moonbows) often appear above waterfalls, where mists of airborne water droplets disperse sunlight into the familiar colours of the visible spectrum: red, orange, yellow, green, blue, indigo and violet. White light is refracted as it enters the droplet, then reflected off the back surface, then re-refracted on the way out at a wide range of angles. A rainbow can be seen only when the sun is directly behind the observer and the airborne droplets are directly in front.

Double rainbow, reflected

Rainbows are complex optical phenomena that come in a variety of forms. Though the single, or primary, bow is the most commonly seen, it is often accompanied by a wider secondary bow, caused by the double reflection of sunlight inside the airborne water droplets. Because the reflection is doubled, the order of colours is reversed in the secondary bow, with red on the inner edge shading to violet on the outside, as can be seen in this serene photograph of a double rainbow reflected in the wet sand of Embleton Bay, Northumberland, England, on a showery June evening in 2002.

Jacob's Ladder

Crepuscular rays are beams of sunlight that are scattered and made visible by minute particles of dust, gas and water present in the lower atmosphere. There are several varieties, including the downward streaming 'Jacob's Ladder', seen here, where the sunlight emerging from gaps in low cloud is scattered by water droplets and divided into columns by the shadows cast by the intervening cloud. The effect takes its name from the episode in *Genesis* 28: 11–17, in which Jacob falls asleep and dreams of angels ascending and descending a ladder connecting Earth to Heaven. This luminous photograph was taken in Northumberland at the end of a rainy day in January 2003. The photographer, Roger Coulam, recalls pulling his car over to take this picture as the sun's rays burst through the cloud layer, 'illuminating the hills and patches of woodland. A simple example of the beauty of the weather.'

Antarctic rays

During a crisp Antarctic sunset in January 1991, spectacular crepuscular rays of the 'Jacob's Ladder' variety (see previous page) illuminate half the sky in a stunning elemental display of water, ice and fiery light. Also known as sun rays or sunbeams, crepuscular rays are so named because they are usually seen during periods of twilight – either at dawn or at dusk – when the contrast between sunlight and shadow is strongest.

Crepuscular rays, type II

The second variety of crepuscular rays takes the form of beams of light that rise upwards from behind a cloud, usually a cumuliform cloud. Like the Jacob's Ladder type, they appear as columns of scattered light separated by shadows that seem to radiate upwards from a single point in the sky. The light rays are in fact near parallel, but because of linear perspective they appear to originate from a particular point-source. The effect is comparable to looking at parallel railway lines, which appear to converge in the near distance. The image on page 83, taken from space, tells the other side of the story.

Anticrepuscular rays

Anticrepuscular rays form in much the same way as crepuscular rays, but appear on the opposite side of the sky, seeming to converge at a position directly opposite the sun, known as the 'antisolar point'. Again, this is a trick of perspective, and the rays are in fact parallel or near-parallel beams, streaming into the curved upper atmosphere. They can be seen only when the sky is particularly clear – as it was on the coast of Florida when this photograph was taken in July 2011 – since any significant dust or haze will reduce the contrast between the shadows and the sky.

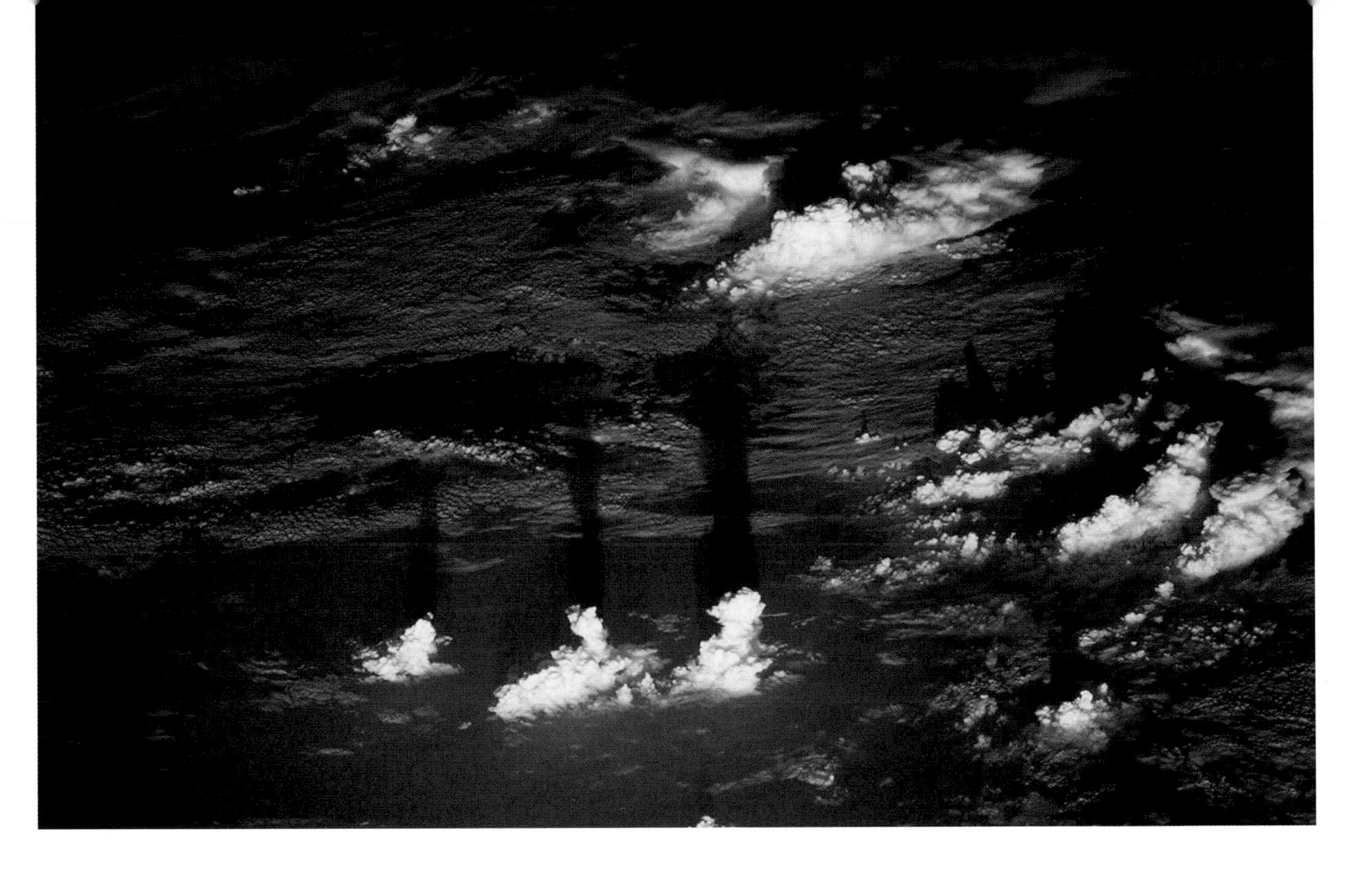

Cloud shadows

This photograph, which was taken by an astronaut on board the now-retired Space Shuttle *Discovery* in June 1998, shows shadows cast by the tops of cumulus clouds in the light of the setting sun. The shuttle was directly over Poland at the time, en route to the *Mir* space station.

From this perspective, the shadows can be seen to be near parallel, whereas from ground level, inside the Earth's curved atmosphere, they would have appeared as radial displays of crepuscular (and possibly anticrepuscular) rays fanning out across the evening sky.

Mirage

A mirage is an optical effect caused by the refraction (bending) of light rays on or near the horizon. Because layers of cold and warm air have different refractive indexes – they bend light at different angles – light rays that pass through sharp temperature gradients will be distorted in interesting ways. In this impressive example, photographed in the Canadian Arctic, pack ice on the horizon appears taller and nearer than it really is. This wall effect, known as a Fata Morgana (after the legendary Arthurian sorceress Morgan le Fay) is a variety of 'superior' mirage, as distinct from the 'inferior' variety, in which the mirage appears below its true position.

Red mammatus

A lurid display of mammatus clouds, lit blood red by the setting sun after a stormy day in the American Midwest. Mammatus clouds tend to form along the bases of extended cumulonimbus anvils, as pockets of cold air sink through the cloud to form these characteristic cellular globes and bulges (see pages 107 and 121 for other examples). They are usually associated with stormy weather, but they can also put in an appearance once things have calmed down, as in this magnificent example of an eerie post-cyclonic sunset.

Cloud iridescence

This beautiful display of pastel-shaded cloud is the result of a process known as irisation: the diffraction, or bending, of light by rows of uniformly sized ice crystals or supercooled water droplets within a thin layer of cloud – in this case a frozen fragment of pileus (or cap cloud) that has formed above a cumulonimbus storm system in central Oklahoma, USA. Formed by strong updraughts within unstable storm clouds, the appearance of pileus indicates that the parent cloud is growing rapidly and is a reliable sign that severe weather is on its way.

Oklahoma sunset

This impressive, glowing evening sky features a complex array of meteorological phenomena in the eerie calm after a day of storms (May 14, 1977). First, the bulbous globes of mammatus clouds loom down from the underside of a dissipating storm cloud's anvil: mammatus form when pockets of cold air sink suddenly from the main part of the anvil. Lower down in the sky are fragments of decaying cumulus cloud, breaking down fast in the cool evening air, while below them is a display of crepuscular rays that appear to fan out from the darkening horizon, west of the town of Hobart, Oklahoma, USA.

Evening sun pillar

This kaleidoscopic image features an array of competing optical effects. On the right a vertical sun pillar rises several kilometres above the setting sun, the result of light reflected from the horizontal surfaces of ice crystals that are slowly falling from the high cirriform clouds. Immediately above the pillar is a bright patch of iridescent cloud, in which the last flare of sunlight has been diffracted into coronal pastel colours. To the left of the image, meanwhile, is a parhelion (from the Greek for 'beside the sun'), a halo phenomenon caused by the refraction of sunlight by ice crystals within the high clouds. Parhelia, also known as 'sun dogs', tend to occur when the sun is low, as in this fine winter sky, photographed from the summit of Flat Top Mountain, North Carolina, USA, on a cold February evening.

Glory and Brocken spectre, Golden Gate Bridge

A glory is an optical phenomenon that appears as a sequence of coloured rings, often surrounding a magnified shadow – the so-called 'Brocken spectre', named after the peak in the Harz Mountains of northern Germany where the phenomenon was first recorded in the 18th century. The effect is caused by light scattered back towards its source by a cloud of uniformly sized water droplets. Because they are cast by low sunlight on to the upper surfaces of clouds or mists, glories are usually seen only by mountaineers or aircraft passengers, but in this case the photographer, Mila Zinkova, was lucky enough to encounter one on a bank of fog beneath the Golden Gate Bridge in San Francisco, California, USA. Note how the glory's concentric rings – the result of the diffraction of sunlight – are centred on the observer's own head, directly opposite the sun at the antisolar point.

Alaskan sun dog

Sun dogs, or parhelia, are optical effects caused by the refraction of sunlight through ice crystals in the atmosphere. They tend to occur when the sun is low in the sky, as in this close-up example photographed on a November evening in an Alaskan forest, in which the sun lies close to the horizon, just out of frame. Parhelia appear to the side of the sun at the same height above the horizon, and often exhibit a range of rainbow colours, with the red end of the spectrum positioned nearest the sun. For this reason they are sometimes known as 'winter rainbows' in northern areas, although, being refractive, ice-crystal phenomena, they have nothing to do with actual rainbows (see page 76).

South Pole halo

Haloes are optical phenomena caused by the refraction of light through tiny hexagonal ice crystals. They often appear in conjunction with other optical effects, as in this superb example, photographed in the icy air above the South Pole on December 21, 1980. The photographer, Lieutenant Cindy McFee of the NOAA Corps, positioned a member of the Antarctic research team in front of the sun to allow the full range of halo phenomena to be seen more clearly. In the centre of the display is a classic 22-degree halo, with a wing-shaped upper tangent arc above it and a pair of bright parhelia, or sun dogs, on either side. Intersecting these is a faint line of light known as a parhelic circle, yet another refraction phenomenon, which appears to pass horizontally through the sun.

Hurricane at sunset

The sunlit serenity of this pastel-hued image is eerily deceptive, for what it shows is the fast-moving turbulence of a powerful Atlantic hurricane; the billowing eyewall at the centre of the image is where the strongest winds are found (see page 28). Category 5 hurricanes (such as Hurricane Katrina at the height of its power) have eyewall wind speeds of around 280km/h (175mph), into which the United States Air Force and NOAA routinely send their manned research planes – the celebrated 'Hurricane Hunters' – in order to monitor hurricane behaviour and growth.

Aurora over the South Pole Observatory

This haunting view of the *aurora australis* – the southern counterpart to the *aurora borealis* (the Northern Lights) – was taken during the winter of 1979 by John C. Bortniak, during his year in charge of NOAA's South Pole Observatory Station in Antarctica. The long winter months on the base, where outside temperatures fall as low as –76°C/–105°F, are enlivened by regular auroral displays, produced by fast-moving particles from the streaming solar wind as they collide with the upper atmosphere; the impact causes atmospheric gas molecules to fluoresce in a range of dramatic colours, sending ribbons of light dancing across the sky.

The green flash

The green flash is a fleeting optical effect that can occur at sunset, just as the last portion of the solar disc dips beneath the horizon. It is caused by the varied refraction of sunlight by the lower atmosphere, which, acting as a giant lens, bends the light from the setting sun towards Earth. In the evening, because of the low angle of the sun, aerosols in the atmosphere tend to scatter the sunlight towards the red end of the spectrum. However, under certain mirage-like conditions, differently coloured images of the setting sun become separated vertically into bands; it is then, if we are very lucky, that we may see a final momentary flash of green as the sun disappears from view.

BROOKLANDS COLLEGE LIBRARY
WEYBRIDGE, SURREY KT13 8T

EXTRAORDINARY WEATHER
STRANGE PHENOMENA

From unexplained holes in the sky to outbreaks of volcanic lightning, this section looks at some of the more surprising displays that are staged by our restless atmosphere.

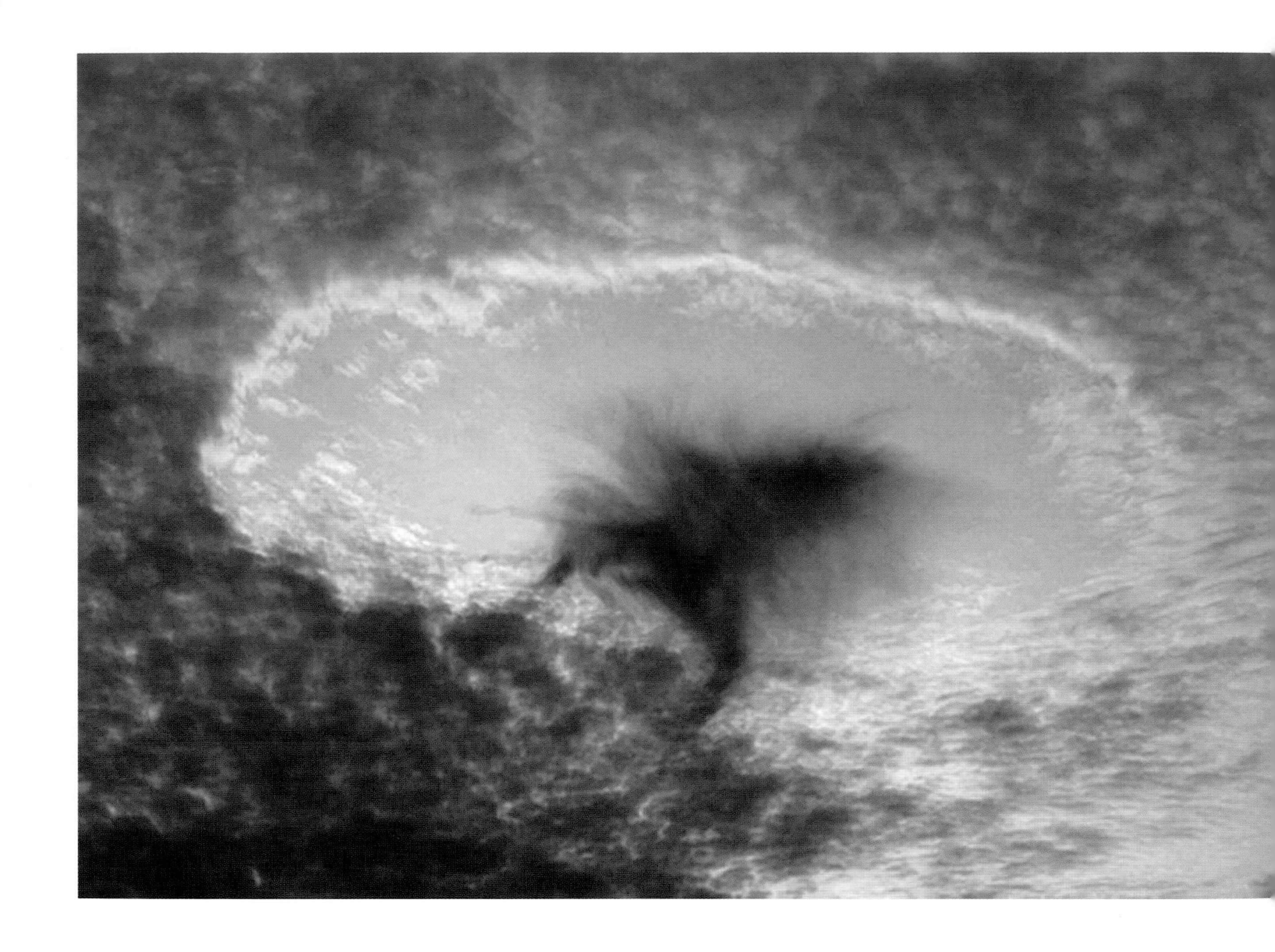

Hole punch cloud

Fallstreak holes – also known as hole punch clouds – occur when patches of high cloud (typically altocumulus or cirrocumulus) freeze and fall away as ice crystals, leaving noticeable gaps in the cloud layer. Their cause remains incompletely understood, though they seem to happen only in supercooled clouds, where water droplets remain in liquid form even at subzero temperatures. It is likely that aircraft exhaust plays some part in their formation, since supercooling occurs when there are not enough 'freezing nuclei' available to initiate the formation of ice from the airborne water droplets. The microscopic particulates from aircraft exhaust are an abundant source of freezing nuclei, so fallstreak holes such as this one – appearing in an altocumulus layer over Austria in August 2008 – may well turn out to be yet another example of anthropogenic, or man-made, weather.

Fallstreak hole

Another dramatic example of a fallstreak hole, this time within a patch of supercooled cirrocumulus cloud suspended high over the Mackenzie Basin, in New Zealand's South Island. This photograph, taken in May 2006, shows the pristine clarity of the atmosphere, the absence of airborne freezing nuclei such as dust, pollen grains or sea salt keeping the water content of the cloud in a liquid or semi-liquid state even in extremely cold temperatures. Research published in the journal *Science* in July 2011 described fallstreak holes as a form of inadvertent cloud seeding (see page 129), produced by the spontaneous freezing of cloud droplets as air flows around aircraft propeller tips or over jet aircraft wings. Such holes tend to form when the plane is climbing or descending through the cloud at a shallow angle, leaving a kind of shadow in its wake.

Fiery fallstreaks

A band of altocumulus clouds turns a flaming red colour as the sun sets over Dallas County, Texas, USA. Note how the fallstreak holes in these clouds are beginning to jettison their icy contents into the lower atmosphere. These streaks will not reach the ground, however, because in atmospheric conditions such as these, falling ice crystals do not melt into liquid raindrops, but sublimate instead – changing their state directly from a solid to a gas – as they pass through the warmer layers of air, evaporating into the fiery evening sky.

Billowing mammatus clouds

A blanket of bruised-looking mammatus clouds bubbles downwards from a turbulent sky as a severe June thunderstorm approaches Woodward, Oklahoma, USA. Mammatus clouds – named from the Latin word for 'udders' – form beneath the anvils of large cumulonimbus clouds. They are caused by the sudden sinking of pockets of cold air from the upper to the lower parts of the cloud, reversing the usual pattern of summer cloud formation in which warm, moisture-laden air convects upwards. They are usually associated with stormy, unstable conditions, although they can also appear some time after a storm has passed.

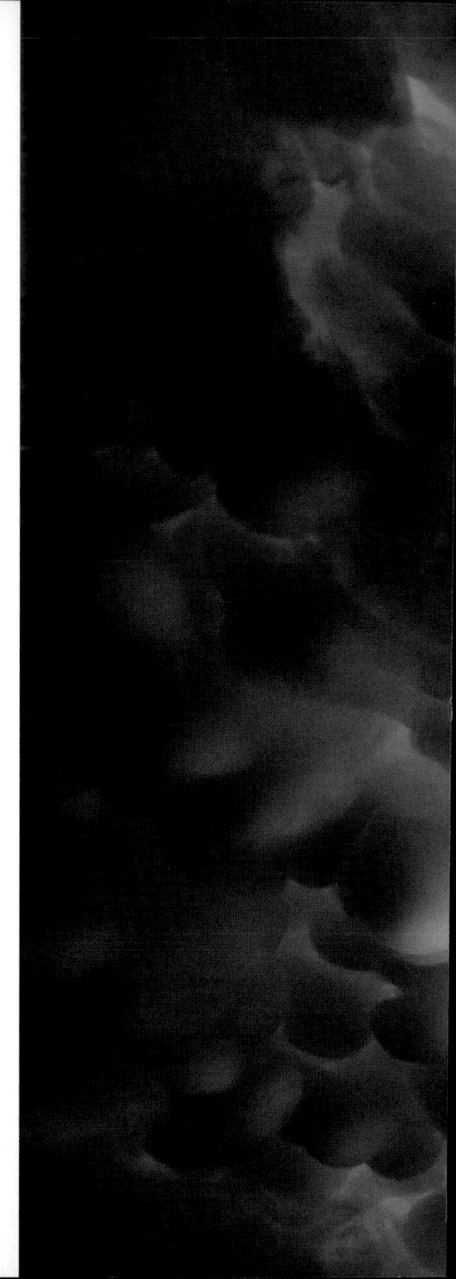

Volcano weather

On April 14, 2010, Eyjafjallajökull (a subglacial volcano in southern Iceland) erupted explosively, jettisoning millions of tons of volcanic ash into the Earth's upper atmosphere. Because the eruption coincided with an unusual high-pressure weather situation across northern and central Europe, weak wind flow meant that the ash cloud remained concentrated as it drifted over the continent, leading to the closure of international airspace for several days. As can be seen from this deceptively serene-looking image, captured by NASA's Terra satellite on April 19, the ash plume followed the curve of the clouds as it drifted southeast, carried by the gently prevailing winds.

Volcanic lightning

In dramatic contrast to the calm of the previous image, this time-exposed photograph, taken on April 18, 2010, tells the other side of the story. Explosive eruptions like that of Eyjafjallajökull are often accompanied by spectacular displays of lightning in and around the ash column, caused by the rapid build-up of static electricity as the rising dust and smoke mixes with the turbulent atmosphere. Pyrocumulus clouds have also formed above this column, as water vapour released by the volcano suddenly condenses on the billions of specks of dust that are churning around in the heated air, creating what meteorologists aptly call a 'dirty thunderstorm'.

Dragon-shaped cloud over Antarctica

On January 28, 2002, NASA's Terra satellite captured this extraordinary image of a cloud formation over the Ross Sea, Antarctica. Was it a dragon? A fish? Or merely an interesting example of the atmospheric physics of convection? The 'eye' of the cloud-creature appears to have been an isolated region of high ice crystals, while the rest of the 'body' was a cumuliform cloud made up of liquid water droplets. It is likely that the distinctive shape of the head was formed when a warm, rising air mass punched through a passing cumuliform cloud. As a convective parcel of warm air rises through the atmosphere, it pushes the colder air out of its way. That cold air spills down over the sides of the rising air mass, and, as it does so, cuts away part of the cloud below, creating (in this case) an accidental work of weather art.

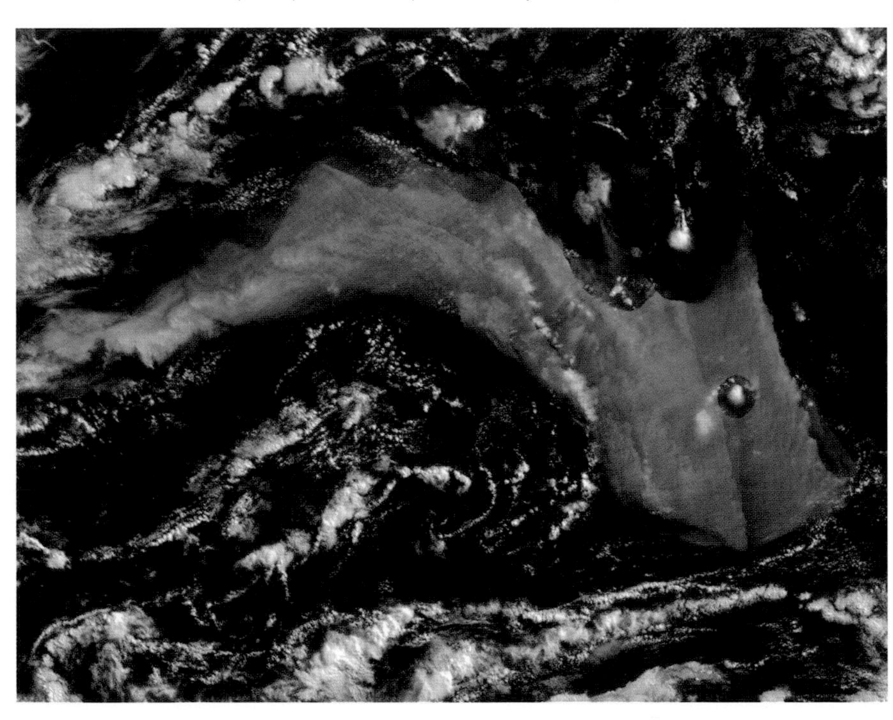

Aleutian clouds

These strange, sponge-like cloud formations were photographed in June 2000 by Landsat 7's Enhanced Thematic Mapper Plus (ETM+) sensor, around 700km (435 miles) over the western Aleutian Islands in the northern Pacific Ocean. Their colour variations are most likely due to differences in surface temperature as well as in the sizes of water droplets present in different areas of cloud. When small droplets merge with others to form larger droplets – a process known as accumulation – the spaces between them grow wider; because wider spaces within a cloud allow more light to penetrate and be absorbed rather than being reflected, those parts of the cloud appear darker.

South Sandwich Island cloud wakes

The South Sandwich Islands are a chain of remote volcanic peaks in the southern Atlantic Ocean that are assailed by year-round westerly winds. Their peaks are high enough to disturb the stratiform cloud layers that pass overhead, creating a train of distinctive cloud patterns known as wakes. The higher the peak, the more substantial the wake pattern it will create, as can be seen in this example, where the taller island in the middle of the image – the 1,005m (3,297ft) Visokoi Island – has generated a boldly defined wake pattern. Further downstream, where the wakes fan out, they merge to form more complex interference patterns. This image was captured in January 2004 by the Moderate Resolution Imaging Spectroradiometer (MODIS), flying aboard NASA's Aqua satellite.

Cloud vortices, South Korea

On a cold day in early March, persistent northwesterly winds blow across the volcanic island province of Cheju-do, South Korea. Mount Halla, the island's tallest peak at 1,950m (6,398ft), blocks a region of flowing air, setting in train a series of rolling spirals, known as von Kármán vortices, in which a sinuous pattern of turbulent flow develops many kilometres downwind from the obstruction. When such turbulence flows through clouds, the resulting paisley swirls can be spectacular, as seen in this image, captured in 2011 by the MODIS device aboard NASA's Terra satellite, from an altitude of around 700km (435 miles).

UFO, New Zealand-style

This impressive lenticular cloud stack hovers low over a New Zealand mountain range at sunset, producing a spectacular golden-orange flying saucer. Such clouds are formed when layers of moist air rise and fall over an obstacle (such as a mountain), creating standing waves of visible cloud in the lee of the peak, where the flowing layers of moisture briefly condense. Like most standing wave clouds, this one hangs at a slight incline because of the downward flow of the leeside air, which only increases its resemblance to a hovering alien craft.

Colorado shelf cloud

A solitary windmill stands sentry as a shelf cloud heralds the approach of a severe June thunderstorm along the east Colorado plains. Shelf clouds appear when pockets of rain-cooled air descend and move ahead of the approaching storm, ploughing beneath the layer of warmer air above to form a low-lying shelf of turbulent, wind-sculpted cloud that (unlike the roll cloud on page 118) remains connected to the base of its parent thundercloud.

Roll cloud, Uruguay

Roll clouds form in much the same way as shelf clouds (see previous pages), except that, unlike shelf clouds, they become entirely detached from their parent thunderstorms. A roll cloud is created when a rapidly sinking air mass hits the ground ahead of an approaching storm, sending a wave of air gusting some distance from the storm itself. This cold flowing air will then slide beneath a layer of warmer air that is being drawn into the storm's vertical updraught, condensing as it does so into a roll of cloud that can end up several kilometres long, like this fine example seen coming in off the Atlantic Ocean along the Uruguayan coast in January 2009.

Tornado and rainbow with hail

On the afternoon of June 12, 2004, storm chaser Eric Nguyen was driving at speed towards the town of Mulvane, Kansas, USA, as a tornado started to develop from a complex of powerful thunderstorms. Pulling over in front of a white farmhouse, Nguyen quickly set up his camera and took a sequence of dramatic photographs that are considered by many storm chasers to be the finest tornado pictures of all time. 'The lighting, colours, and the storm were all absolutely perfect!' he recalled, and the way the tornado, rainbow and cloud base appear to intersect at a point above the trees adds a certain geometric beauty to this image. (The white streaks in the sky are hailstones.) The tornado went on to cause significant damage to property, and later, as it began to decay, debris – including roofing insulation – fell from the sky like snow.

Mammatus descending

The unmistakable looming shapes of mammatus clouds (see page 106 for an account of their formation) can be glimpsed above a ragged deck of decaying altocumulus cloudlets. This spectacular sunset, with its unusual, multilayered tapestry of colours, shapes and textures, appeared over Fannin County, Texas, USA, at the end of an early summer's day that had been marked by severe thunderstorms.

Microburst

A microburst is a sudden descent of a column of cold air from the high base of a raincloud, producing strong, destructive winds at ground level. Microbursts come in two varieties – wet and dry – depending on whether or not they are accompanied by falling rain. The wet microburst seen here is hurtling towards the ground to the left of a tall, vertical column of rain-darkened cloud. Such downbursts can be powerful enough to flatten trees and farm buildings, and they pose a significant danger to aircraft during takeoff and landing. A number of fatal plane crashes in the United States have been attributed to microbursts over airports.

Asperatus clouds, Oklahoma

A turbulent variety of altostratus undulatus cloud, asperatus clouds appear in eerie, wavelike formations, often in the aftermath of convective thunderstorm activity. The new term 'asperatus' (which derives from the Latin for 'rough') has only recently been proposed to the World Meteorological Organization, whose members are currently considering whether or not to add it to the official meteorological lexicon. If they decide in its favour, it will be the first new cloud term to be added since 'intortus' ('twisted') was adopted in 1951. The term 'asperatus' was originally proposed by Gavin Pretor-Pinney, founder of the Cloud Appreciation Society.

Kelvin-Helmholtz waves

Kelvin-Helmholtz ('K-H') waves (or billows) are rare, short-lived formations that occur when the boundary between a warm air mass and a layer of colder air beneath it is disturbed by strong horizontal winds, causing the upper layer to move faster than the lower layer. Just as the wind excites waves on the surface of the sea, this aerial turbulence causes the 'crests' of the waves to move ahead of the main body of the cloud, leading to the characteristic wave formations seen in this photograph, taken in the Rocky Mountain National Park in northeast Colorado, USA.

EXTRAORDINARY WEATHER

MAN-MADE WEATHER

Human influence is just as evident in Earth's atmosphere as it is on the ground. This section looks at an array of man-made effects, from cloud seeding to artificial lightning, and asks whether our weather is getting weirder and, if so, has it anything to do with us?

Man-made lightning

Ever since Benjamin Franklin flew a kite into a thundercloud in 1752, scientists have been fascinated by the creation of artificial lightning. In 2008, man-made lightning within a thundercloud was triggered for the first time by physicists at the Langmuir Laboratory for Atmospheric Research in New Mexico, USA. They used remote laser pulses to create ionized channels of air molecules that triggered and conducted the charge. In this more modest example of lightning research, cloud-to-ground lightning has been generated by firing a small rocket into a thundercloud. A long piece of fine copper wire trailing from the rocket provides an easy path for the lightning to follow to reach the ground. Such research allows scientists to measure the current, voltage and other parameters of lightning bolts, and is frequently used to test the capacity of new kinds of equipment to survive an electrical strike.

Cloud seeding

Devised in the 1940s, cloud seeding is a weather modification technique that remains in surprisingly widespread use. It involves sprinkling the upper layers of rain-bearing clouds with stimulants such as dry ice or silver iodide aerosols, in order to encourage supercooled water droplets to freeze into ice crystals, thereby hastening and intensifying the natural rainmaking process. Most cloud seeding is done for irrigation purposes in arid farming regions such as Western Australia, but storm and hail mitigation has also been attempted using cloud-seeding techniques. Project Stormfury, for example, was a United States Government-funded initiative that ran from 1962–83, with the aim of learning how to weaken hurricanes by stimulating the freezing of supercooled water within the energetic parts of the cloud: all attempts proved unsuccessful. This photograph was taken on September 13, 1969, during one of the project's research missions.

Contrails above northern Virginia

Contrails – short for 'condensation trails' – are now among the most familiar clouds in our skies. Also known as vapour trails, they are high, icy clouds, formed from the water vapour and solid particulates emitted from aircraft at altitudes above 8km (around 26,000ft), where the air temperature is well below freezing. Their lifespan and behaviour act as a visual guide to the wider atmospheric situation. If the upper air is particularly warm or dry, then no contrails will form, or they will be very short-lived. If it is already laden with supercooled moisture, the contents of an aircraft's exhaust will precipitate rapid contrail formation, giving rise to extensive, criss-crossing layers of spreading cloud, as in this photograph taken by NASA scientist Louis Nguyen in January 2001.

Contrails over Newfoundland

A variety of clouds can be seen in this satellite image of the sky above Newfoundland, Canada, from the cumuliform (top right) to the cirriform (bottom right), with a skein of man-made contrails in the middle that show how busy the flight corridors between Europe and North America have become. Like natural cirrus clouds, contrails will thicken and spread in advance of an approaching weather front, sometimes growing to cover enormous areas of sky. So far, the longterm climatic effects of such man-made cloud layers have yet to be firmly established (see opposite).

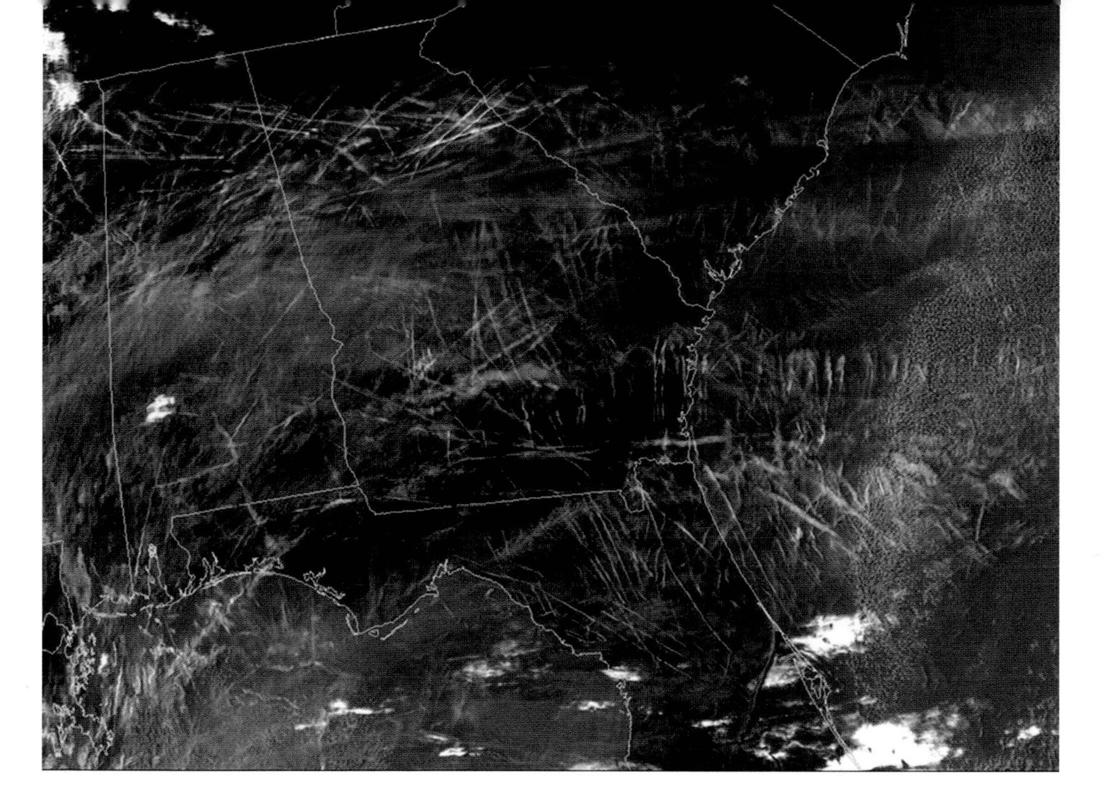

Contrails over the southeastern United States

The increase in aviation over the past half century has transformed the contrail into the world's most abundant cloud type, though the longterm implications of this remain little understood. Data from the contrail-free skies that followed the 9/11 attacks showed slightly warmer days and slightly cooler nights than usual, caused by more sunshine reaching the ground by day and more radiation escaping at night. Icy contrails are especially persistent during the winter months, as can be seen from this enhanced infrared image from NASA's Terra satellite, which shows widespread contrail layers formed at different altitudes over the southeastern United States during the morning of January 29, 2004. Recent research suggests that a rise in night flights may be responsible for a slight increase in night-time warming, caused by the prevalence of contrails over the world's teeming flight paths.

Ship tracks over the northeast Pacific

The maritime equivalent of aircraft contrails, ship tracks are linear clouds that form around the aerosol particulates from the smokestacks of larger vessels. As can be seen in this image, captured from NASA's Terra satellite on March 4, 2009, ship tracks appear brighter than the natural marine clouds around them, because of the smaller size and greater abundance of the droplets they contain. Cloud droplets form when water vapour condenses on to small particles, such as those found in exhaust aerosols; the more particles there are, the more droplets there will be within the cloud, and the droplets will be correspondingly smaller. Because smaller droplets have a greater reflective index, the ship track clouds appear brighter than natural clouds, and bounce more sunlight back into space: so could the deliberate production of such artificial marine clouds prove a viable method of combatting future warming?

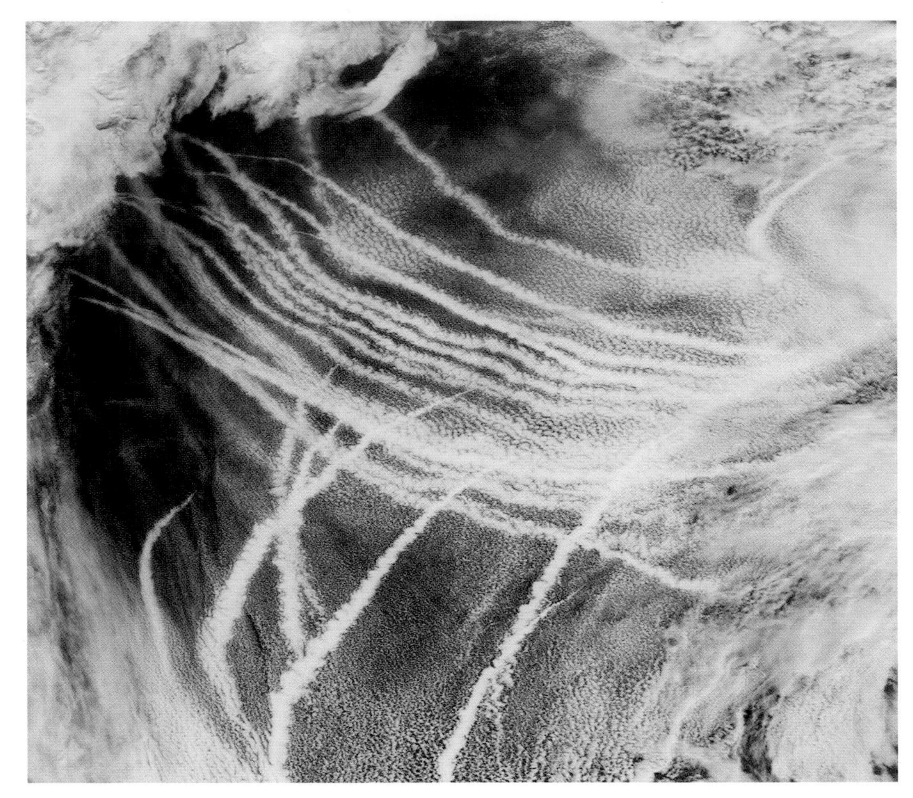

Fallstreaks and canal clouds above Texas

As was seen on page 104, when aircraft pass through layers of cloud they leave visible gaps and streaks in their wake; long linear gaps are known as distrails (short for 'dissipation trails') or canal clouds, a number of which can be seen from above in this image captured by NASA's Terra satellite on January 29, 2007, as it passed over Texas and Louisiana, USA. The holes in this layer of stratocumulus cloud were made by a variety of civil and military planes taking off or landing at Dallas/Fort Worth International Airport as they rose or descended through the cloud. This inadvertent form of cloud seeding could be responsible for increased winter snowfall near major airports such as DFW, although on this occasion most of the jettisoned ice crystals evaporated long before they reached the ground.

Wingtip vortex

Wingtip vortices are cylinders of low-pressure, turbulent air that form in the wake of accelerating aircraft. Their cores can sometimes become visible as high-altitude ribbons of cloud, owing to the sudden condensation of water vapour at very low pressure. In this example, however, photographed at NASA's Wallops Flight Facility in eastern Virginia, USA, the vortex created by the wing of a crop-dusting plane has been artificially visualized at ground level through a layer of coloured smoke. Such wake vortices can be surprisingly powerful and long-lived, affecting the stability of nearby aircraft, especially during takeoff and landing. They are a challenge to air traffic controllers, who must choreograph flight departure and arrival times around these hazardous man-made wind tunnels.

Stratospheric rocket contrail

At 8.56am on May 16, 2011, the manned Space Shuttle *Endeavour* blasted off on its final mission, delivering parts to the International Space Station. This photograph, taken from a shuttle training aircraft, shows the rocket's icy contrail spearing up into the stratosphere above the Kennedy Space Center in Florida, USA, as it headed towards its destination more than 300km (185 miles) above the Earth. The Space Shuttles' exhaust plumes were 97 per cent water, and condensed in the upper atmosphere during the eight minutes they took to reach orbit in the thermosphere. Rocket contrails do not remain vertical for long, but are blown by the winds into sinuous shapes before evaporating on their way back down to Earth.

Noctilucent clouds over Estonia

Noctilucent clouds (NLCs) are the highest clouds in Earth's atmosphere, forming on the fringes of space some 80km (50 miles) up in the mesosphere. They were first observed and named in the 1880s, at the height of the Industrial Revolution, when they were considered to be the rarest of clouds, visible only from polar regions. Nowadays they not only appear more often than before, but also shine brighter and are observable from increasingly lower latitudes. Why is this? One hypothesis points to the fact that extreme cold – around –130°C (–202°F) – is needed to form icy clouds in environments as dry as the mesosphere, and if the mesosphere grows colder as the lower atmosphere warms, then more NLCs are likely to be formed. (Mesospheric cooling occurs when surface radiation is trapped by Earth's warming atmosphere and is thereby prevented from making its way out to space.)

Noctilucent clouds from space

If it transpires that the observed increase in noctilucent cloud formation is indeed due to mesospheric cooling, and these clouds' increased brightness is due to larger ice crystals being formed from a high-altitude influx of water vapour from the warming layers below, then the visible impact of human activity will have extended much farther into the atmosphere than was previously imagined. We will, it seems, have succeeded in spreading man-made weather not just over the skies above us, but out towards the very edge of space.

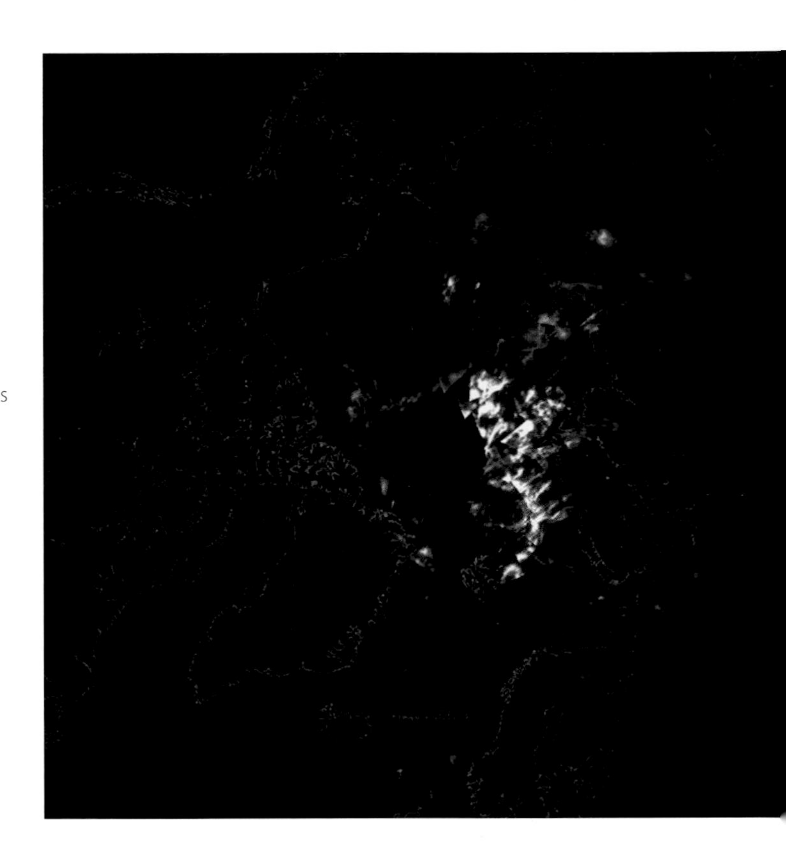

Index

Picture Credits

Alamy © p37 (Gerth Roland/Prisma Bildagentur AG); Corbis UK Ltd. © pp18 (Mike Hollingshead/Science Faction), 51 (Michael Freeman), 60 (Arctic-Images), 119 (Eric Nguyen); NASA © pp26/27; 83; 137; 139 (Cloud Imaging and Particle Size Experiment data processing team at the University of Colorado Laboratory for Atmospheric and Space Physics); NASA - Goddard Space Flight Center (GSFC) © pp32; 36 (Jeff Schmaltz/MODIS Rapid Response Team), 44 (SeaWiFS Project/ORBIMAGE), 53 (SeaWiFS Project/ORBIMAGE), 108 (Modis Rapid Response Team), 110 (Jacques Descloitres/MODIS Land Rapid Response Team), 111 (USGS EROS Data Center Satellite Systems Branch), 112 (Jacques Descloitres/MODIS Rapid Response Team), 113 (Jeff Schmaltz/MODIS Rapid Response Team), 132 (SeaWiFS Project/ORBIMAGE), 134 (Jeff Schmaltz/MODIS Rapid Response Team), 135 (Jeff Schmaltz/MODIS Rapid Response Team); NASA - Langley © pp131 (Louis Nguyen), 133; 136; NASA - Lyndon B. Johnson Space Center © pp13; 14/15; 52; 56/57; National Meteorological Library © pp40/41 (Matthew Clark), 43; 76 (Andy Best), 80 (Keith Shone); National Ocean and Atmospheric Administration (NOAA) © pp3/4; 12 (C. True/AOML/Hurricane Research Division), 21 (NOAA Photo Library/NOAA Central Library/OAR/ERL/National Severe Storms Laboratory), 22 (F.N. Robertson), 25 (NOAA Photo Library/NOAA Central Library/OAR/ERL/National Severe Storms Laboratory), 28; 29 (Steve Nicklas/Royal Air Force Photograph), 39; 42 (NOAA Photo Library/NOAA Central Library/OAR/ERL/National Severe Storms Laboratory), 45 (Collection of Dr. Herbert Kroehl/NGDC/Bill Koch), 50 (NOAA George E. Marsh Album), 59 (Dan Borsum/NWS/WR/WFO/Billings Montana), 64/65; 68; 69 (Shane Lear), 79 (Dave Mobley/Jet Propulsion Laboratory), 89 (NOAA Photo Library/NOAA Central Library/OAR/ERL/National Severe Storms Laboratory), 90 (Grant W. Goodge), 94 (Lieutenant (j.g.) Cindy McFee/NOAA Corps), 95; 96/97 (Commander John Bortniak/NOAA Corps), 122 (NOAA Legacy Photo ERL/WPL), 124 (Collection of Dr. Bill Hooke), 129 (AOML/Hurricane Research Division); Roger Coulam © pp72; 77; 78; Science Photo Library © pp20, (Jim Reed), 31 (Fred K. Smith), 38 (Dr Juerg Alean), 66/67 (G. Antonio Milani), 73 (Sam Walton), 84/85 (Louise Murray), 93 (Michael Giannechini), 99 (Pekka Parviainen), 109 (Olivier Vandeginste), 128 (Peter Menzel); U.S. Department of Defence © p54 (Cpl. Alicia M. Garcia); Weatherpix Stock Images © pp16; 17 (Gene Rhoden), 23 (Gene Rhoden), 24 (Gene Rhoden), 33 (Gene Rhoden), 55; 70/71; 74/75; 86; 87 (Gene Rhoden), 105 (Gene Rhoden), 106/107 (Gene Rhoden), 115; 116/117 (Gene Rhoden), 121 (Gene Rhoden), 123 (Gene Rhoden); Wikipedia Commons © pp1 (Daniela Mirner Eberl), 46 (Kathrin Spiegler/Salvi 5), 47 (Petr Dlouhy), 61; 82 (Wojtow), 92 (Mila Zinkova), 102 (H. Raab), 104 (Dollsworth), 118 (Daniela Mirner Eberl), 138 (Martin Koitmäe).

Cover image © Science Photo Library

Back cover image © Martin Koitmäe

These images have come from many sources and acknowledgment has been made wherever possible. If images have been used without due credit or acknowledgment, through no fault of our own, apologies are offered. If notified, the publisher will be pleased to rectify any errors or omissions in future editions.

Further Reading

Burroughs, William (ed), *Climate: Into the 21st Century* (Cambridge, 2003)

Burt, Christopher C., *Extreme Weather: A Guide and Record Book* (New York, 2004)

Craig, Diana, *The Weather Book: Why It Happens and Where it Comes From* (London, 2009)

Dunlop, Storm, *Weather: Spectacular Images of the World's Extraordinary Climate* (London, 2006)

Eden, Philip, *Change in the Weather: Weather Extremes and the British Climate* (London, 2006)

Fish, Michael et al, *Storm Force: Britain's Wildest Weather* (Ilkley, 2007)

Hamblyn, Richard, *The Cloud Book: How to Understand the Skies* (Newton Abbot, 2008)

—, *Extraordinary Clouds: Skies of the Unexpected from the Beautiful to the Bizarre* (Newton Abbot, 2009)

Herd, Tim, *Kaleidoscope Sky* (New York, 2007)

Higgins, Gordon, *Weather World: Photographing the Global Spectacle* (Newton Abbot, 2007)

Hollingshead, Mike, and Eric Nguyen, *Adventures in Tornado Alley: The Storm Chasers* (London, 2008)

Jennings, Steve, *Time's Bitter Flood: Trends in the number of reported natural disasters* (Oxford, 2011)

Lynch, John, *Wild Weather* (London, 2002)

Met Office, *The Met Office Book of the British Weather* (Newton Abbot, 2010)

Mogil, H. Michael, *Extreme Weather* (London, 2007)

Reed, Jim, *Storm Chaser: A Photographer's Journey* (New York, 2007)

Robinson, Andrew, *Earthshock: Hurricanes, Volcanoes, Earthquakes, Tornadoes and other Forces of Nature*, 2nd edn (London, 2002)

Simons, Paul, *Weird Weather* (London, 1996)

—, *Since Records Began: The Highs and Lows of Britain's Weather* (London, 2008)

Acknowledgments

It is a pleasure to thank all who helped in the production of this book, especially Neil Baber, Stephen Burt, Roger Coulam, Freya Dangerfield, Beverley Jollands, Hannah Kelly, Gavin Pretor-Pinney, Adrian Simpson, and the many photographers whose work is featured here. I would also like to thank the librarians and archivists of the National Meteorological Archive, The British Library, the University of London Library, and the Science Museum Library.

A DAVID & CHARLES BOOK,
© F&W Media International, Ltd 2012

David & Charles is an imprint of F&W Media International, Ltd
Brunel House, Forde Close, Newton Abbot, TQ12 4PU, UK

F&W Media International, Ltd is a subsidiary of F+W Media, Inc.,
10151 Carver Road, Cincinnati OH45242, USA

Text and designs copyright © RIchard Hamblyn 2012
Layout and Photography © David & Charles 2012, except those listed on page 142

First published in the UK and US in 2012
Digital edition published in 2012

Layout of digital editions may vary depending on reader hardware and display settings.

Richard Hamblyn has asserted the right to be identified as author of this work in accordance with the Copyright, Designs and Patents Act, 1988.

All rights reserved. No part of this publication may be reproduced in any form or by any means, electronic or mechanical, by photocopying, recording or otherwise, without prior permission in writing from the publisher.

A catalogue record for this book is available from the British Library.

ISBN-13: 978-1-4463-0191-3 Paperback
ISBN-10: 1-4463-0191-5 Paperback

ISBN-13: 978-1-4463-5623-4 e-pub
ISBN-10: 1-4463-5623-X e-pub

ISBN-13: 978-1-4463-5622-7 PDF
ISBN-10: 1-4463-5622-1 PDF

10 9 8 7 6 5 4 3 2 1

Acquisitions Editor: Neil Baber
Assistant Editor: Hannah Kelly
Project Editor: Beverley Jollands
Senior Designer: Jodie Lystor
Production Manager: Beverley Richardson

Paperback edition printed in China by RR Donnelley for:
F&W Media International, Ltd
Brunel House, Forde Close, Newton Abbot, TQ12 4PU, UK

F+W Media publishes high quality books on a wide range of subjects.
For more great book ideas visit: www.fwmedia.co.uk